Tectonic
Globaloney

Tectonic Globaloney

N. Christian Smoot

Copyright © 2004 by N. Christian Smoot.

Library of Congress Number: 2003098607
ISBN: Hardcover 1-4134-3729-X
Softcover 1-4134-3728-1

All rights reserved. No part of this book may be reproduced or transmitted in any form or by any means, electronic or mechanical, including photocopying, recording, or by any information storage and retrieval system, without permission in writing from the copyright owner.

This book was printed in the United States of America.

To order additional copies of this book, contact:
Xlibris Corporation
1-888-795-4274
www.Xlibris.com
Orders@Xlibris.com

Contents

1. Introduction 11
2. Background 13
3. Geodynamic Models Culminate
   with the Plate Tectonic Model 17
4. Applications of the Plate Tectonic Model 26
5. Neglect of Field Data at the Altar
   of Conceptual Models 31
6. Field Data vs. Globaloney 38
7. So, What Do We Have Here? 134
8. Gaia's Basic Forces: An Updated Working Hypothesis ... 139
9. Applications 142
10. Tectonic Globaloney:
    Your Tax Dollars Hardly at Work 160
Acknowledgements 163
Appendix I: Professional Background: 165

Dedication

I dedicate this work to the field hands who gathered and processed the information. The multi-beam sonar data are from the declassified US Naval Oceanographic Office Ocean Survey Program files. The single-beam and earthquake data are from US National Geophysical Data Center files. The satellite altimetry data are from the GEOSAT Program of the US Government. Many rock ages are from Deep Sea Drilling/Ocean Drilling Program files; many are from other sources. The tectonics portion is dedicated to the inquisitive scientists who cared enough to evaluate the data on the table and seek to derive a more meaningful explanation of Earth geodynamics.

globaloney – word used by students when they are being sold shares of the Brooklyn Bridge by snake-oil salesmen. It means just what it says: "I don't believe you."

1

Introduction

THIS IS A booklet about things I learned during my career as a deep-ocean surveyor/marine geomorphologist, a career which spanned 1966-1998, with a few years off for good behavior in the mid-1970s. Essentially my job was to survey select portions of the Atlantic and Pacific ocean floors, with the Caribbean, Mediterranean, and South China Seas thrown in for good measure (see Appendix), for the nation's submarine fleet. They used the survey data as "road maps." Along the way, sometime during the early 1970s, the Navy hierarchy decided to start releasing some of our classified bathymetry to the public. I think I was the only one left in-house during lunch one day when the boss went looking for a poor, unsuspecting person to be the conveyor of that information.

Anyhow, I was selected to release all of our information on guyots. After scratching my head and wondering for a while, I called my old bud Peter Vogt at the Global Ocean Floor Analysis and Research (GOFAR) Division of the US Naval Oceanographic Office (NAVOCEANO) and asked him what a guyot was. I had worked with Peter on a project for the Ocean Survey Program (OSP), of which I was a part.

Well, Peter gave me the scoop, and I started reading up on the subject. Plate tectonics was a new kid on the block, so I had not been exposed to it at Clemson during my geology course, so I had to learn about that too. Vogt was/is an expert on that subject.

A few papers into this, and I went to a symposium in 1981 at Texas A&M. I showed a feature called the Michelson Ridge, a rather large feature that I had a part in surveying, and that contained four guyots. Oh, yes, they are flat-topped seamounts that were theoretically eroded subaerially before subsiding back below sealevel. *The guyots, in fact the entire ridge, seemed anomalous to what few facts I had gleaned from the literature. However, Don Hussong, Hawaii Institute of Geophysics, reviewed my paper and told me that, because I was the first to show these features, I was allowed to take the first cut at interpreting their history.* That is where the tectonics comes in. Geomorphology is what you get when you apply geodynamics over time to any feature.

Given that encouragement, and given the fact that we were the only surveyors using a multi-beam sonar system, I began a long career of writing papers and books. What I learned from my work and writing was that all was not understood what should have been understood before certain parts of the plate tectonic hypothesis were formulated, and what follows is a summary of that. Don't get nervous, the science is kept to a minimum for the lay reader. Cites are mostly not even listed, although they are in most of the papers in the Appendix. After having lived in Hawaii for a few years, I have learned the value of "talking story," and that is the manner in which this is presented – with a large helping of aloha.

2

Background

CHARLES WYNN AND Arthur Wiggins used Sidney Harris' cartoons to apprise us of their most significant scientific breakthroughs (*The Five Biggest Ideas in Science*, 1997): (1) the model of the atom presented by physics, (2) chemistry's periodic table, which sorts the elements, (3) astronomy's Big Bang theory, concerning the origination of the universe, (4) the plate tectonics model of geology and geophysics to explain Earth evolution, and (5) Darwin's theory of evolution, a biology adjunct used to explain the living world. This list seems fairly comprehensive. Not being a chemist, astronomer, or astrologist, the fourth item seemed the most attractive to a one-time practicing ocean floor geomorphologist. And, I also realize that in this expanding body of knowledge we enjoy today, possibly some of the basic premises may be a little outdated. After all, the idea took hold about 1966 or so.

Right off the bat, the audience needs to be established because the intent is not here in trying to break through the hardened shell of the practicing geodynamicists, a group whose paradigm was predetermined in 1966. The interest lies in getting some of the facts out to the students and practicing neophytes who are still able to think outside the box, assimilate real data/knowledge, and formulate their own thinking without modifications being implanted by preconceived notions. This is about Earth geodynamics, the evolution of the current working hypotheses, and the foibles and fables of dealing with a concept that requires the fitting of real data into the theory rather than the other way around. An attempt shall be made to unravel the esoteric mysteries, explain what has happened to it along the way, a life history from birth to death as it were, and also to try to derive a replacement mechanism by which students of Earth sciences may see a way to explain geodynamic events in a more robust manner without having to rely solely on *ad hoc* sideroads at every aberration; an attempt to grasp the epitome of the situation so to speak.

Figure 1. Equatorial belt on Venus having all the appearance of having been hydrodynamically created. We will see this same feature again on Mars and on Earth.

All science has to begin with some basic premises, and so shall we. Earth geodynamics deals with Mother Earth, which formed about 4.55 Ga (billion years ago). Many hypotheses have put us here, not the least of which is the "Big Bang." I don't know; I wasn't there. Along the way, many forces have acted upon the planet. Earth resides in an orbit about the sun between the planets Mars and Venus. Mars and Venus are classified as "dead" planets because no life exists on them, and because no visible tectonic activity seems to be taking place at the present time. They are cold planets. Obviously, life and tectonics both exist on Earth. Interestingly, both Venus and Mars appear to have some sort of equatorial belt, a belt which would seem to be braided and splayed (Figure 1); it looks almost like a delta. This type of geomorphology will be seen again on Earth, later in this presentation. Please remember it. Additionally, more is known about the surfaces of these two sister planets than is known about Earth's surface, believe it or not.

Inroads are being made through several avenues. The deep hole drillsite in Kola, Siberia has taken us down 12 km into Earth's center. Inferences have been drawn from several fields, not the least of which are earthquake seismology and seismic stratigraphy, as a means of trying to decipher what is going on underneath us. The speed of sound through the earth varies with the substrate being sounded. Sand could carry an 800 ft/sec speed, and unfractured granite could carry one of 20,000 ft/sec. From this we may say that the denser the material, the faster the speed of sound, and this is the first conclusion to be considered for the working hypothesis.

Two schools of thought are used to explain what is thought about Earth's center: (1) Earth's core is hot, liquid rock or (2) it is cold, liquid plasma. The hot, liquid core is proposed to be comprised of iron and nickel, and in this model the Earth is a "heat engine." Rising from the core, this turns into a slushy inner mantle, some spots/lines of warmer material called the asthenosphere, a cooler outer mantle, and a solid outer crust. Everything above the asthenosphere is lumped into the "lithosphere." Two types of crust exist, continental and oceanic. Oceanic crust consists of basalts, which are primarily silicon dioxide (like sand), a fine-grained, iron-rich mafic rock. Continental crust consists of granites, which are lighter than basalt. Granite is a coarse grained, iron-poor felsic rock. Both are igneous rocks that form under different regimes. As we saw above, the speed of sound through granite is fast, about 20,000 ft/sec.

On the other hand, the cold, liquid core is proposed to be a mixture of hydrogen and helium, constantly transforming into atoms, the last of which to form is iron in the outer core. Everything above the asthenosphere remains the same as the first school of thought. This is a relatively new concept, and it will be discussed further. We may draw the conclusion here that the speed of sound through a liquid is not as fast as through granite.

Additionally, Earth has three possibilities, just as do we; it can expand, contract, or maintain the status quo through some means.

Before dealing with real-life data relating to geodynamics, the reader needs to have a basic understanding of the evolution and background on the derivation of the plate tectonic hypothesis. To that end, the basic tenet of earth science as we know it in 2003, the plate tectonic model, seems to have captivated most of the monies being sunk into geodynamic research at the present time. An introduction is in order.

3

Geodynamic Models Culminate with the Plate Tectonic Model

TECTONICS IS THE key to unlocking the structural geometry. Lithosphere motions determine regional structure. These motions are either horizontal or vertical, such as the sinking of a basin or the rising of a mountain range. Lithosphere motion after the principle production phase gives rise to additional stresses, which may or may not change the geomorphology of an existing feature. Secondary and tertiary tectonics create secondary and tertiary features on the primary structure. Thus, the geomorphology may be used to determine the tectonic history of a particular region, or the entire Earth for that matter, and the geomorphology of discrete pieces of the ocean floor are the primary topic herein.

Historically, the Ice Ages were thought to have caused much of the surface geomorphology in Europe. That was when scientists began to drift away from the "flood" and try to decipher what they were really finding. This included the discovery of fossilized remains of dinosaur bones and plants, which were older than Bishop Ussher's 5000 or so years.

James Hutton, a Scottish geologist, was instrumental in developing the age. In an article from *Encyclopedia Britannica*: "At that time geology in any proper sense of the term did not exist. Mineralogy, however, had made considerable progress. But Hutton had conceived larger ideas than were entertained by the mineralogists of his day. He desired to trace back the origin of the various minerals and rocks, and thus to arrive at some clear understanding of the history of the earth. For many years he continued to study the subject. At last, in the spring of the year 1785, he communicated his views to the recently established Royal Society of Edinburgh in a paper entitled "Theory of the Earth, or an Investigation of the Laws Observable in the Composition, Dissolution and Restoration of Land upon the Globe." Hutton expounded that geology is not cosmogony, but must confine itself to the study of the materials of the earth; that everywhere evidence may be seen that the present rocks of the earth's surface have been in great part formed out of the waste of older rocks; that these materials having been laid down under the sea were there consolidated under great pressure, and were subsequently disrupted and upheaved by the expansive power of subterranean heat; that during these convulsions veins and masses of molten rock were injected into the rents of the dislocated strata; that every portion of the upraised land, as soon as exposed to the atmosphere, is subject to decay; and that this decay must tend to advance until the whole of the land has been worn away and laid down on the sea-floor, whence future upheavals will once more raise the consolidated sediments into new land. In some of these broad and bold generalizations Hutton was anticipated by the Italian geologists; but to him belongs the credit of having first perceived their mutual relations, and combined them in a luminous coherent theory based upon observation."

Next came Charles Lyell, another Scots geologist. In 1828 he explored the volcanic region of the Auvergne, then went to Mount Etna to gather supporting evidence for a theory of geology he was developing, that of uniformitarianism. In essence, given sufficient time, millions of years, geological change was slow and gradual and not subject to inexplicable catastrophe such as Noah's Flood. In 1829, volume one of his great work *Principles of Geology* appeared.

Not to bore you with a bunch of stuffy definitions, but a few will help the average reader to get more into the subject matter:

anticline – Geologic structure in which the rock layers have been formed
    into an arch. Erosion exposes the oldest rocks at the axis of an
    anticline.

fault – A fracture or break in Earth's lithosphere along which differential movement of the rock masses has occurred.

foot wall – The side of the fault containing the rock layers that are below the fault plane.

geosyncline – Continental margin downwarping into Earth's crust that has seen sedimentaion and volcanic activity.

hanging wall – The side of the fault containing the rock layers that are above the fault plane.

mobile belt – Long, relatively narrow region of tectonic activity (past or present) that delimit cratons. Mobile belts contain a thick, mostly complete column of strata, but the rock record is hard to read because of folding, faulting, unconformities, metamorphism and igneous intrusion.

normal fault – A fault in which the hanging wall has moved down relative to the foot wall.

reverse fault – A fault in which the foot wall has moved down relative to the hanging wall.

strike-slip fault – A fault in which the displacement along the fault is horizontal.

structural basin – Geologic structure in which the rock layers dip in toward the center of the structure, usually in concentric circles. Erosion exposes the oldest rocks at the edges of the structure.

syncline – Geologic structure in which the rock layers have been formed into a trough. Erosion exposes the oldest rocks at the edges of a syncline.

From these small beginnings, things like geosynclines, anticlines, miogeosynclines, and that sort found their way into the earth scientist's vocabulary. Tectonism is characterized by orogenesis or geosynclinal activity. Vertical tectonism became the order of the day, with erosion wearing down mountains, giant sedimentary basins forming, and isostatic compensation lifting mountains up again through thrust faulting. This could be described as "pulsation without representation" on a global scale. The rising and falling of the land allowed for the incursion by the sea, so that we had shallow oceans at one time over the mid-western United States, as an example, where the desert and high plains now reigns supreme. That is why we can find fish skeletons in Wyoming. This was all very well understood by the early 1900s. Rather than merely pulsating and maintaining the same size, by the 1930s some were suspecting that Earth was actually expanding; that is, Mother Earth is getting a middle-age spread.

However, these explanations did not explain earthquakes, volcanoes, deep ocean trenches, and the apparent fit of some of the continents. Something was missing, and the geologists could not quite pinpoint the solution.

Then Alfred Wegener (1915), a meteorologist, is given credit for deriving the continental drift hypothesis in his landmark book, *The Origin of Continents and Oceans*. He noticed that the eastern South American and western African continental margins seemed to fit together, so he arbitrarily placed them in close juxtaposition in the distant past and derived a new hypothesis. This idea was relegated to the "crackpot bin," where it lay for many years before being incorporated into a later hypothesis. Kiyoo Wadati (1927) first plotted the deep earthquakes around Japan. With the beginnings of earthquake seismology, B. Gutenberg and C.F. Richter predicted that Earth is a series of plates separated by active seismic belts (1949). M.L. Hill and T.W. Diblee, Jr. noted large horizontal displacements on Earth's surface along great faults (1953). Hugo Benioff deciphered the deep earthquake zones (about 650 km) around the Pacific as being interpreted as great thrust faults (1949). Bill Menard discovered (1955) the large fracture zones in the North Pacific. Ron Mason verified fracture zones by the extensive magnetic patterns on the ocean floor that ended abruptly with his work on the Pioneer in 1955. Soon after this, Keith Runcorn (1956) used magnetic pole displacement based on 180 million-year-old (Ma) rock samples to show that North America had been displaced from Europe, and this figure became the age of the oldest seafloor extant. The great rift system of the midocean ridges showed the dynamic state of the ocean floor by Bruce Heezen in 1960. Harry Hess theorized seafloor spreading in his by-now famous paper, *History of Ocean Basins*, in 1962. Alan Cox, R.R. Doell, and Brent Dalrymple (1963) developed a paleomagnetic time scale using a mass spectrometer. Fred Vine and Drummond Matthews (1963) expanded on the earlier studies of Mason's magnetic anomalies. A dude named Morley came up with the same idea at that time, but he is only lately being given a share of the credit. Tuzo Wilson (1965) showed that the magnetic anomalies were offset on formation along transform faults instead of after the magnetic signature had been imprinted.

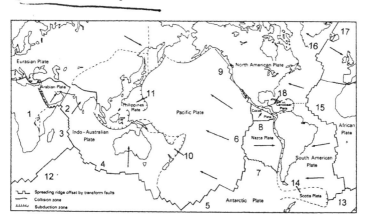

Figure 2. Plates defined by divergent (MOR) and convergent (subduction zone and collision margin) boundaries defined by linear bands of earthquakes where: 1=African Rift, 2=Carlsberg Ridge, 3=Mid-Indian Ridge, 4=Southeast Indian Ridge, 5=Pacific–Antarctic Ridge, 6=East Pacific Rise, 7=Chile Rise, 8=Galapagos/Carnegie Ridge, 9=Gorda, Endeavor, Juan de Fuca ridges, 10=Lau-Havre Ridge, 11=Mariana Trough, 12=Southwest Indian Ridge, 13=Atlantic-Indian Ridge, 14=Scotia Arc, 15=Mid-Atlantic Ridge, 16=Reykjanes Ridge, 17=Mohns Ridge, 18=Cayman Trough.

The transform faults proved to be the key, and the tectonic revolution was underway. At the 1966 Geological Society of America meeting in San Francisco, Lynn Sykes proved Wilson's hypothesis by studying earthquake motion, and Fred Vine tied all of the preceding together. The deep earthquakes and great thrust faults became the descending plate at the subduction zones. The remelt of the leading edge of that plate rose through the lithosphere in liquid (magma) form to pour out on the surface and become lava for volcanoes. The remelted lithosphere (magma) flowed on a great conveyor belt back to the midocean ridges, and there it rose to the surface to create new lithosphere. This point is recognized as the start of the plate-tectonic revolution.

The plate tectonic model was more-or-less finalized in 1966. Generally, the meeting of the minds revealed a fairly simple conveyor-belt explanation, and this is the gist of horizontal tectonics. Midocean ridges, or spreading centers, are zones of shallow seismicity where magma wells up to the surface from the asthenosphere, forms new oceanic crust, and moves off-ridge, all the while cooling and subsiding. This makes the spreading center a divergent plate boundary. The rate of movement varies but is measured in centimeters per year (cm/yr). The crust moves and ages with distance away from the midocean ridge crest. That crust eventually, about 180 Ma, is subducted into the upper mantle in a region known variously as the subduction zone, the Wadati – Benioff zone, or the convergence zone or margin. The subduction zone is characterized by deep earthquakes and is also a convergent plate boundary. Continental and oceanic volcanic arcs form landward to Wadati-Benioff zones and are the foci of extremely virulent volcano and earthquake activity. The Pacific "ring-of-fire" is an excellent example of this phenomenon. Transform faults and fracture zones interconnect the midocean ridges and subduction zones. They are the foci of shallow earthquakes and are sometimes thought to be plate boundaries. Features not associated with those means of crustal production are generally attributed to hot spots, which Jason Morgan (1972) hypothesized to be fixed diapirs centered in the mantle. Through time, the ridges, transforms, and trenches may move about in no organized manner producing such phenomena as plate reorganizations, changes in plate movement direction, polar wandering, magnetic shifts, trench migration, and others. We will not delve too far into that, as the subject is well-covered elsewhere.

Hailed by the Earth geodynamic community as the universal panacea, the plate – tectonic hypothesis was adopted *in toto*. The new hypothesis answered many questions about the origins and functions of the midocean ridges, continental rifts, fracture zones, volcano production and active arcs, deep sea trenches and the attendant Benioff zones, ophiolites, accreted melanges, etc. that had never had a satisfactory explanation.

Although they are tainted by many reinterpretations, the original ideas about wandering continents were included, now called the "Bullard fit." The drifting of continents was proven by, among other things, the existence of a Triassic tetrapod fossil called *Lystrosaurus*. *Lystrosaurus* had been found in India, South Africa, and Antarctica, thus allowing the juxtaposition of all of those continents during the lifetime of *Lystrosaurus*. That cluster of continents including South America was called Gondwanaland, a supercontinent. *Lystrosaurus* was used to prove continental drift because these continents, while being joined during the Triassic Era (about 225 million years ago (225 Ma), are no longer joined. Within the plate-tectonic continental configuration, another supercontinent existed to the north, Laurasia. The extreme edges of Gondwanaland and Laurasia were in the cold oceans, and the interiors bordered the warm Tethyan Sea. Naturally, no *Lystrosaurus* could be found on both because it could not get across such a wide body of water or such a climate extreme as must be crossed under the constraints of that continental configuration.

How did the continents wander? you may rightly ask. As noted, Earth was predicted to be a series of lithospheric plates that were separated by active seismic belts very early in the game. In the oceans, the crust is about seven km thick, while the lithosphere, which includes the crust and upper mantle, is about 80 km thick. The continental cratons are about 600 km thick, and they ride around on the oceanic plates. The plate boundaries are classified as divergent and convergent. The divergent boundaries are the midocean ridges (Figure 2). The convergent boundaries are subduction or collision zones where plate materials either are consumed, accreted, or built upwardly (Figure 3). Newer fabric may be created there as the compressional forces buckle the meeting plates somewhat. Finally, the plates can slide past each other in a strike-slip action, leaving parallel-ridge and trough structures. The plates join, or weld, at different times to other plate segments, which means that they react differently at different times. The rigid mass of lithosphere, or plate, moves over a spherical surface, Earth. Of necessity, some form of random pressure must release both internally and externally to this rock. That release is in the form of fracturing, expressed bathymetrically as fracture zones by proper name. The fracture zones are Earth's great cooling cracks, and they point in the direction of plate motion. In a hypothesis that employs the principles of a conveyor belt system, everything that happens while the belt is moving affects the geomorphology.

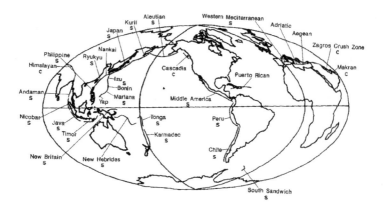

Figure 3. Convergent margins of the world where "s" is subduction zones and "c" is collision margins.

Several types of tectonics are at play here. Relative motion of the lithosphere, or plate; that is, relative to the deep upper mantle, is shown by the spreading phenomena such as fracture zone azimuths and the ocean floor fabric. Absolute motion of the lithosphere; that is, relative to the surface, could be defined by the linear features, such as the seamount/island chains, or megatrends. This means that two sets of azimuths could co-exist on the same lithosphere, and that absolute motion could change randomly, such as the proposed 43 Ma event. Because of these two sets of trends, relative motion cannot be shown for the Pacific Ocean basin by the fracture zones; they will cross each other somewhere in the western basin. It should be obvious by now that, if absolute plate motion occurs anywhere, it must also occur everywhere else at the same time. For the plate tectonic hypothesis, you see, requires that

Earth maintain a constant diameter, and the plates are solid rock. Push and pull is the name of the game.

As of this writing 12 larger plates are recognized (Figure 2): North American, South American, Pacific (Figure 15), Eurasian, African (Figure 23), Indian-Australian (Figure 24), Philippine (Figure 26), Antarctic, Caribbean, Scotia, Cocos (Figure 20), and Nazca. The smaller microplates in the plate-tectonics hypothesis are called the Mariana, Adriatic, Arabian, Aegean, Juan de Fuca, Bismark, Solomon, Fiji, Magellan, Manihiki, Gorda, and several others. A movement is afoot to increase this number of plates almost daily as the geophysicists give us more and newer interpretations, the latest being the Indian-Australian plate split.

And, we still are no closer to being able to predict earthquakes, volcanoes, and other deleterious, at least to us, phenomena.

4

Applications of the Plate Tectonic Model

FOR THE WORKING hypothesis to function, it must explain every geologic phenomenon in every instance. Otherwise, we are left in the netherworld of *ad hoc* explanations for every aberration, wouldn't you say (and, plate tectonics is the ultimate repository for adhockism!). Considering the fact that the ocean floor, covering over 70% of Earth's surface, was largely unexplored in 1966, the fact that any sort of working hypothesis evolved is nothing short of a miracle in itself. The plate tectonic hypothesis was then compiled by a bunch of seers, or a bunch of people who cared little for the facts. We'll see as we follow these examples.

Figure 4. Continental drift and seafloor spreading, defined by Chris Scotese and a group of friends, for the past 600 Ma. By

200 Ma the continental configuration called Pangaea had started to break apart to begin their drift towards today's alignment.

Using that basic premise, the continental drift idea has been used in a variety of scenarios. About 660 Ma everything came together to for a supercontinent called Grenville. Its initial opening phase allowed the Iapetus Ocean to form, separating Euroamerica and Gondwanaland. Iapetus lasted from 770 (Proterozoic) to 225 (Triassic) Ma. The ocean basins began forming by 545 (Cambrian) Ma when Grenville started breaking up (Figure 4). A full description of plate tectonics at work is in many book and papers that I alone have been associated with (see Appendix), not to mention the folks who really do this for a living. Ergo, any expansion on that interpretation would be superfluous.

A modification to Figure 4 combines the work of Ian Dalziel, Bill Thomas, Ricardo Astini, Eldridge Moores, and Victor Ramos: North America has been shown to be responsible for the emplacement of the exotic terrane in the Andes. The two continents were joined at 545 Ma. Then, as North America pulled away from South America 515 Ma, the Ouachita terrane broke free, crossed the Iapetus Ocean 475 Ma, and rejoined South America 465 Ma. The aftermath of this is that North America ran around the north end of South America and faced off against Africa 322 Ma, and collided with both continents about 265 Ma. During the Permian Era, Laurasia and Gondwanaland were united to form Pangaea, giving Earth one massive land body surrounded by the Panthallasan Ocean (eo-Pacific). All of this started breaking up about 200 Ma (Triassic Era), with the Tethys Sea forming between them.

The oldest ocean basin, the Pacific, has existed for more than 600 Ma. However, attempts to find such dates have been futile. Celal Sengor, a self-professed armchair geologist, had the hubris to design a model of eastern Eurasia (Figure 5), a model in which he did not consider any of the available geology from the Himalayan region. Between 610-530 Ma, Russia and Angara were one, faced on the east by the Turkestan Ocean. From 430-424 Ma, Russia split off and was free-wheeling in a clockwise direction while Angara was rotating counter-clockwise, creating the Kipchak Arc. At 390-386 Ma, Russia re-docked with the Kipchak Arc to form an enclosed Khanty-Mansi Ocean. The whole shooting match was then drifting clockwise. At 367-362 Ma, the Khanty-Mansi Ocean was being pinched as a large sea was created on the Turkestan Ocean side. From 332-318 Ma, the Khanty-Mansi Ocean continued closing as Russia rotated more into Angara. As Russia rotated counter-clockwise, the subduction system closed more. Between 318-303 Ma, the process was complete, and the Khanti-Mansi Ocean was totally closed. For future reference, the Khanti-Mansi Ocean is called the Ural Ocean, as it underlies the Ural Mountains in this model. The Turkestan Ocean is the Panthalassan, and the subduction zone in this model is the present western Pacific trench system. The Kipchak Arc is presumably Kamchatka, Japan, Korea, and eastern China.

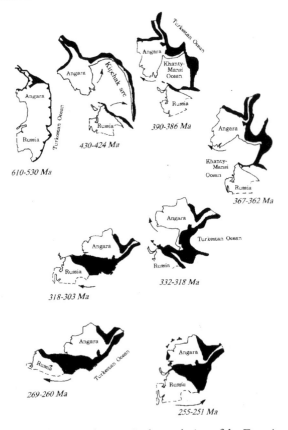

Figure 5. Theoretical events in the evolution of the Eurasian landmass from 610-251 Ma (based on ideas of Celal Sengor.).

Angara is the nucleus of Siberia, and Russia is the nucleus of Europe (1). The Turkestan Ocean floor subducts, causing a chain of volcanoes to rise landward of that. Angara and Russia split apart (2), creating a new ocean and two new subduction zones. Angara rotates clockwise and Russia moves north (3, 4 and 5). Between 350-320 Ma the Tarim block approaches from the south and collides. Angara and Russia rejoin (6), closing the Khanty-Mansi Ocean. Russia slides to the west (7) as the South China block collides from the southeast. By 250 Ma Russia swings back again (!), and the Altaid region is assembled. Asia continues to grow as a succession of continental blocks and island arcs collide, the most notable of which is India.

Names; whoever derives the equation gets to name the parts. Part of science, you know. You could do the same thing and label it all "blue widgets." Somebody will publish it! The plate tectonicists immediately gravitated towards this idea, accepting it totally and without question.

I participated in a conference at Texas A&M University in 1983, the topic of discussion being subduction zones. I took the bathymetry of the Michelson Ridge with me. The feature contains a plateau and four guyots, and it is over 500-km long. It had beautiful flow structures pouring out eastwardly, and the ridge is on an E-W azimuth. Beautiful. As the co-convener, Zvi Ben-Avraham, was wrapping things up, he noted that the conference was good for one thing; everyone there had finally gotten to see some real ocean floor data. Naturally, this made me feel good, being able to contribute something from our declassified archives to help the civilians/academia. It was a shame that this was 17 years after the formulation of the PT hypothesis!

The oldest portion of the Pacific plate is proposed to be the part adjacent to the engulfment zone, or trenches on the NW portion of the basin. The oldest borehole to date has recovered 156 Ma material (ODP Site 801). Lithosphere flexure gives ages of 170-171 Ma for 12.8°N latitude, 156.8°E longitude and 21.5°N latitude, 159.2°E longitude. The Pigafetta Basin is placed at 160-170 Ma. A transect across a subduction zone has never been drilled down to basalt in the entire history of the Deep Sea Drilling Project (DSDP) and its successor, the Ocean Drilling

Program (ODP). Leg 60 at 18°N latitude attempted, but was not successful. Leg 125 also failed at 31°N latitude. Oceanic lithosphere is a 10-km thick and generally uniform shell, which is mostly basalt. As a last resort, earth scientists have used the magnetic stripes found ubiquitously. The older stripes are called the "M" series. The ages listed for the "M" series are not ground – truthed, so that measurement is merely used to show a chronology in relative ages, as in "this" happened before "that." Additionally, the current magnetic maps of the ocean basins include the catch-all, non-descriptive terms "Cretaceous Magnetic Quiet Zone" and "Jurassic Magnetic Quiet Zone." This is because nobody had the data to make the determination as to the ages of the ocean floor in those regions.

Each portion has been subjected to a myriad of explanations as to "where" and "why." The proof of this was related mostly to the magnetic stripes. The western Pacific basin is largely labeled the "Jurassic Magnetic Quiet Zone" and the "Cretaceous Magnetic Quiet Zone." With no data, any interpretation you want can be had, and that is exactly what the earth expansionists want you to believe. They did not, any of them, use this shipboard magnetics data in any of their interpretations.

5

Neglect of Field Data at the Altar of Conceptual Models

BY THE MID-1970s everyone more-or-less accepted the plate-tectonic hypothesis. What? You don't believe that? Funny, neither did I. Our tax dollars, mucho dinero to be exact, paid for that worthless bit of nonsense. A couple of useful items could have been developed by the working hypothesis so many have come to depend upon. It would have been nice if we had learned how to predict when an earthquake or volcano was going to occur. Whether the plates float around is virtually worthless to the ordinary citizen. The rate of one inch per year average, roughly the speed of your fingernail growth, will not affect any of us in our lifetimes. So what good is tectonics?

A review of alternate hypotheses formulated during this time frame will serve a useful purpose; many of the older concepts are still valuable in the formulation of an updated working paradigm. Most of those older hypotheses contained either Earth expansion or Earth contraction in one form or another, although Earth contraction had been falling out of favor. Synclines, anticlines, geosynclines, eugeosynclines, and other forms of vertical tectonics were used in different situations. Ian McDougall theorized hot lines rather than hot spots (1972). The geologist team of Howard Meyerhoff (father) and Arthur Meyerhoff (son) immediately questioned the validity of the fledgling plate tectonic hypothesis. The Meyerhoffs introduced the idea of heated channels above the asthenosphere (1974). The idea of periodic surges was introduced very early by A. Rice and Rodes Fairbridge (1975). It was called "cyclic thermal runaway" and was predicted to have occurred as recently as 90-100 Ma, during the "Great Cretaceous Outpouring." Lynn Sykes, one of the founding fathers of the plate-tectonic hypothesis, discovered that alkaline magmatism and earthquakes extend for several hundreds of kilometers inland from the bathymetric ends of fracture zones (1978). Sykes also noted that the transforms (fractures) that develop on the opening of an ocean basin undergo an abrupt change in strike on land while displaying a small amount of offset. The Walvis Ridge, Rio Grande Rise, and New England Seamounts in the Atlantic Ocean basin were all proposed to be such sites.

The basic problem is that spurious-to-non-existent bathymetry had been used to prove the plate-tectonic hypothesis. Much better data sets became available in 1974 from NAVOCEANO. In the first place, the fracture zones implanted on the Pacific Ocean crust lay in a fanning pattern (Figure 6), opening to the east across the basin, which would tend to show that the seafloor spreading occurred from a spot on the west, and that the plate grew as it spread outwardly. Additionally, the dating of the ocean floor "basement" has long been suspect. Discrete points on the ocean floor have a wide range in age and have caused a veritable proliferation in a new type of feature, the "micro-plate," such as the Magellan, Kula, Easter, and others. The Global Time Scale has been constantly updated, so the times for previous proposed events will vary from older to more recent studies. The magnetics-based ages for the Pacific plate are in a constant state of fluctuation. Additionally, the pole-of-rotation was found to have moved 8° in the past 60 Ma, 20° in the past 200 Ma, and at least 90° in the past billion years.

A problem with geometry may have been the impetus for new Earth evolution explanations; simply put, the math doesn't work. Even without decent bathymetry of the ocean floor, the midocean ridges, trenches, and collision margins have been known since the mid-1960s. The ridges produce a discrete amount of ocean floor, so the collision margins must necessarily be the avenue of destruction of that same amount of ocean floor. The midocean ridges measure 74,000 km (Figure 2), and this means that 148,000 linear km of new material is constantly being produced according to the current hypothesis. In theory, that much linear distance in collision margins must exist to keep Earth from having a middle-age spread leading to another "Big Bang" situation. There is not; there are only 30,500 km of trenches, about one-fifth the linear distance producing new ocean floor. The Mediterranean-Zagros-Himalayan-Indonesian collision zone to the Timor Trench only adds 9000 km more to the take-up figure (Figure 3), still only one-fourth the amount of spreading ridges. Surely this information leads one to suspect that Earth is expanding.

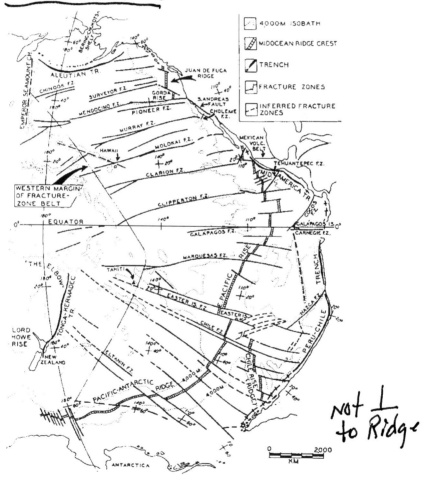

Figure 6. A 1972 attempt at delineating the major features of the Pacific basin. Considering the "fact" that the fracture zone azimuths show the direction of plate motion, and

considering the "fact" that the plate as an entity can only move in one direction at the same time, this diagram should have shown the need for a course correction at the outset. Instead, the plate tectonicists of the era did nothing. As a matter of fact, that august body still does not recognize the fallacy of the original fracture zone definition, even in the presence of the orthogonal intersections.

Geometric problems abound. A plate has to have a spreading center and a convergent margin. Most, unfortunately, leave one wondering where it all went wrong. The realization that the Antarctic plate has no convergent margins whatsoever should have raised a red flag. The African plate is surrounded on the west, south, and east by spreading ridges. The only thing that may be construed as a convergent margin is the Aegean Trench, a rather minuscule feature between Africa and Greece. Similarly, no convergent margin exists for the proposed Eurasian plate (Figure 3). On a smaller scale, no convergent margin can be found for the Gorda, Juan de Fuca, and Cocos plates, although much taxpayer dollars are spent trying to locate a trench at the Cascadia margin off Washington and British Columbia. It does not exist in the bathymetry; it does not exist in the earthquakes. We are wasting our time and money looking there. The Manihiki and Magellan plates are nothing more than upheavals; rises as it were. They do not have either a spreading center or a convergent margin. Lastly, no plate boundary exists between the North and South American plates (see dotted line on Figure 2). The plate adherents can take heart, though, the Pacific, Arabian, and Philippine plates all fit the classic definition.

At the time all of this was taking place, there was one glaring oversight by the excited geoscience community: by 1966 not enough of the ocean floor had been sampled to derive any meaningful explanation about how the ocean floor formed, and that constituted about 70% of Earth's surface. Nobody on Earth knew what the ocean floor looked like, and nobody had any kind of handle on the age of that floor. This includes the bathymetry, rock samples, shipboard magnetics data, and satellite altimetry, all parameters that remained to be sampled in a meaningful relationship during the intervening years. But, they still had an Africa/South America cuddly fit.

Most mobile models hide behind the Pacific Quiet zones to derive any explanation necessary for the fruition of that particular model. However, they make a big mistake. The Jurassic Magnetic Quiet Zone does not exist. According to shipboard magnetics data collected by NAVOCEANO, synthesized by David Handschumacher and Gene Morganthaler in the early 1980s, the Jurassic Magnetic Quiet Zone of the western Pacific actually shows the plate to be moving southeasterly, with the younger portion to the north. If we remain within the working hypothesis, then the western Pacific plate is actually moving southeasterly and subducting into the Vityaz Trench system north of the Coral Sea/Fiji Basin.

Using plate-tectonic constraints and the work of Handschumacher and Morganthaler (Figure 7) gives a Pacific Oceran basin interpretation thusly: Before M38 (170 Ma) a triple junction separated three paleo-plates; the Izanagi on the NW, the Farallon on the east, and the Phoenix on the SW. The Izanagi plate is also called the Bering plate.

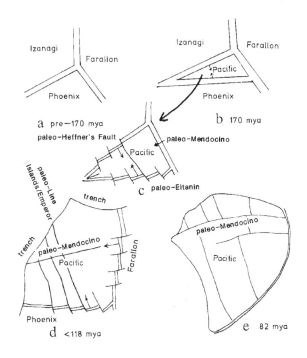

Figure 7. I used the shipboard magnetic information synthesized by Handschumacher and Morganthaler to hypothesize the Pacific "plate"/basin formation.

By M38 a minuscule Pacific plate had started forming at this junction. Fracture zones created by this spreading formed in a NNW-SSE direction, making them the oldest fractures on the present Pacific seafloor. Trenches formed to the NW, and the Izanagi plate subducted. The Phoenix-Pacific spreading center took an active burst, and wheeled to the SE, dragging the pre-existing fractures with it. These would be the present south Pacific WNW-ESE-trending fracture zones. The Phoenix-Pacific Ridge, still joined on the north to the Farallon-Pacific Ridge, began to take the appearance of the present East Pacific Rise by 118 Ma. Fractures formed at the Farallon-Pacific Ridge became the WSW-ENE-trending fracture zones, like the Mendocino, Murray, and other north Pacific fracture zones. This gave the fanning pattern. By 82 Ma, the orthogonal fractures were imprinted, and the acute megatrends began to form.

At 43 Ma a drastic change in motion direction occurred, as witnessed by the "elbows" in several of the continuous seamount/island chains. That remains the motion direction to this day.

The spreading direction from the northern Pacific-Izanagi boundary would have been to the SE on the Pacific plate, creating the first set of fracture zone lineations from the north. The Pacific-Izanagi trunk channel apparently produced crust in that direction from M38 until M1. Coevally, at M38, spreading on the south of the basin started creating new Pacific plate and has continued to the present in a NW-direction from the Pacific – Phoenix trunk channel, now called the Pacific-Antarctic Ridge. The older portion of the NW-trending fracture zones is now at the center of the Pacific plate. Thus, the fracture zones produced by the two spreading centers, the Pacific-Izanagi and the Pacific – Phoenix, may have joined each other or been in close juxtaposition to produce plate-wide N – S megatrends.

The Pacific plate grew rapidly on the west because of opposing spreading, or extension, by the Pacific-Phoenix and Pacific-Izanagi MORs. The Pacific-Farallon spreading center, on the east of the basin, increased activity at about M0. This started the creation of a westward-growing component of the Pacific basin. The modern analog to the non-spreading situation lies in the eastern Pacific. With no spreading centers between central Mexico and the Gulf of Alaska, no extension exists. The San

San

Andreas Fault is an extinct continental margin fracture zone that produces no new extensional features. The magnetic lineations aging westerly from North America are not explained in the plate tectonic framework. They should age toward the north in to the Gulf of Alaska; i.e., parallel to the direction of spreading.

This is the general outline of the formation of the Pacific basin, remaining as I said within the constraints provided by the working hypothesis. As Earth wobbled on its axis, the pole-of-rotation changed. The compression-formed fractures also underwent realignment. Just as the present pole is where it is, so is the formation of the fracture zones indicative of the location of those poles.

In summary, because opposing spreading centers will be creating new lithosphere faster than a lone spreading center, it appears that the combined effort of the Pacific – Izanagi and Pacific-Phoenix spreading has caused the western portion of the Pacific plate to grow appreciably faster than the eastern portion of the basin. That changed the plate's shape from a westward pointing triangle to more rectangular. Fracture zones created by the NNW-SSE contraction have precluded the possibility of any of the eastward-growing Pacific-Farallon fracture zones from crossing the plate until such time as the NNW-SSE trending fractures re-welded. Of passing interest here is that the Pacific "plate" and the African "plate" appear to be mirror images (antipodes) of each other, with the African "plate" being about half the size of the Pacific "plate."

And, while we're here visiting the geophysicists, somewhere during all this action this erudite body decided that Earth's core was a hot, liquid mass. *Theory earth hot liquid mass*

6

Field Data vs. Globaloney

NOT THAT THE original intent was to attack the global tectonic working hypothesis, the idea was actually to diagnose events between the formulation of the hypothesis and the present day to determine a possible course correction, should one be necessary, based on my experience. Along the way it was discovered that all is not as it seems, or some folks want it to be, all the time. In fact, it was discovered that it is seldom as it should be. Not only did we need a course correction but also somebody has changed the entire navigation suite! So, a new and unusual way to do business was opted for; I collected the data and see where it led; like letting the horse pull the cart; or the dog wag the tail. Starting with the most concrete evidence, the rocks, the discourse continues with the bathymetry, probably because that is something you can actually see rather than dwelling in the dream world of geophysics. I spent 30 years of my life doing deep-ocean surveys and compiling the charts at NAVOCEANO (again, see Appendix). Hopefully, this will show that no snake-oil is being peddled here!

Each section begins with the basic premise of the working hypothesis. Then the actual field data is presented. Once the ocean floor age is firmly established, the spreading centers where new ocean floor is created (midocean ridges; Figure 2) comes next, followed by the convergent margins where this ocean floor is returned from whence it came (subduction zones, collision margins; Figure 3), the evidence of seafloor spreading (fracture zones and magma plumes), and the geophysical aspects of same.

6-1. Geology *vs.* Geophysics: a Rocky Beginning

I was told by a not-to-be-named, unimpeachable source along the way about a Caribbean cruise with the late, lamented Bruce Heezen, one of our premier early ocean floor researchers. He was conducting a survey in the Cayman Trough (Caribbean region), needing some fresh basalt to prove that the Trough was actually a spreading center. Heezen grabbed samples for an entire cruise before he finally got one sample of basalt. He based his entire report on that one sample.

A similar story was given by Dong Choi. When he prepared a manuscript on the presence of continental crust and a paleoland during the Paleozoic to Mesozoic in the northwestern Pacific based on all hard evidence, including dredgings, DSDP drillings, paleogeography, seismic data interpretation, and geology of the Japanese Islands, the Chief of his Division ordered not to publish it. The Chief threatened him, "if you publish this paper, you have no room to stay with us", and accused him "you are doing bad science". Of course Dong published it and quit the organization soon after that. The paper appeared in a prestigeous "Journal of Petroleum Geology" in 1987 with a reviewer's comment, "quite revolutionary... almost certainly the precursor of the future in seismic exploration and tectonic interpretation".

And with this epiphany, we begin the study of tectonic globaloney.

Plate Tectonic Presumption: The geophysicists have given us a seafloor age of about 180 Ma based on several magnetic anomalies in the Gulf of Alaska. In 1998, we were treated to this summation by the head of the ODP, Tom Davies: "... rocks of the ocean floor are everywhere much younger than rocks found on continents, indeed we have yet to find rocks older than 200 MY in the deep oceans." In fact, Site 801 produced an age of 150 Ma, the oldest yet reported by the ODP. Additionally, the ocean floor is comprised of a heavier basalt, and a lighter granite-type rock of which the continents are comprised floats around in this sea.

- Mid ocean ridges
 ocean floor is returned from whence it came
 subduction zones, collision margins

From the ODP website: *The Ocean Drilling Program (ODP) is funded by the U.S. National Science Foundation and 22 international partners (JOIDES) to conduct basic research into the history of the ocean basins and the overall nature of the crust beneath the ocean floor using the scientific drill ship JOIDES Resolution. Joint Oceanographic Institutions, Inc. (JOI), a group of 18 U.S. institutions, is the Program Manager. Texas A&M University, College of Geosciences is the Science Operator. Columbia University, Lamont-Doherty Earth Observatory provides Logging Services and administers the Site Survey Data Bank.*

Any opinions, findings, and conclusions or recommendations expressed in these documents are those of the author(s) and do not necessarily reflect the views of the National Science Foundation, the participating agencies, Joint Oceanographic Institutions, Inc., Texas A&M University, or Texas A&M Research Foundation.

Actuality: That may be well for their study area, but the rest of the literate world is also looking at the ocean floor, not just that portion near land. And, the rocks collected from the ocean floor are nowhere near the age, or composition, of what you have been taught.

To begin with, the magnetic values used to establish the 200-Ma age are not ground – truthed in anything; they are hanging in the air. Several scientists have made a study of rocks across the San Andreas Fault and published the results in 1999. The Fault has many folks worried, especially those living in the belt between Los Angeles and San Francisco. You have been told repeatedly that California would end up in Alaska. The "proof" has been a proposed 3000-km of offset on the seaward side during the past several hundred million years, moving Baja California north. This proof is based on magnetic stripes. One group decided that they had bought enough of this snake oil. The researchers took a look at the geology of the region and found very old rocks, called zircons, on both sides of the Fault. Guess what? The displacement was a maximum of 500 km in the past 2 Ga. Not to worry, California will stay right where it is during your lifetime! This fact should have quietened the sensationalist newsmedia, but it has not. We've got to get more of those newspapers and popular magazines into people's homes!

But, what about the ocean floor? Let's get one thing straight right away, and that is the age of the ocean floor. We don't need any magnetics or hypothesized ages, we've got rocks – lots and lots, nay, thousands, of rocks, in fact. Geology, meaning the study of rocks, plays a large part in the derivation of a new hypothesis. Rock samples have been collected by the nationally-funded DSDP and ODP. While not one of these core samples has reached off-ridge, oceanic basement rock in the 35 years of the programs, the tectonicists still use the value assigned to the magnetic anomalies, as well as the ocean floor age, at about 180 Ma. Heaven forbid that anyone would rock that boat.

We are dealing with an equipment problem here. Once enough sediment has "rained down" onto the underlying stratum, heat and pressure take over. The much compacted sediment is metamorphosed into an extremely hard rock called chert. If one were to take the time to read the 1300-some-odd hole descriptions, one would immediately see that most boreholes stop at this layer because the drill bit will not penetrate chert. So, we are instead given extrapolations. "We couldn't get through the chert, but we figure the age and composition of the layer beneath that is no older than . . ." You get the picture.

Well, much to the collective dismay of these drillers, others have collected rock samples, not under the aegis of the ODP. Their samples tell us an entirely different story, one that proves the ocean floor is an order of magnitude older than the 180 Ma age proposed so early by the plate-tectonicists and adhered to by the expansionists. Many of these rocks have been explained away by such statements as "ice-rafting" (on the equator!); "ship's ballast dumping" (on the summit of Tahiti!); and glacial movement (as in moraines in the central Atlantic Ocean basin). The community could feel the squeeze being applied by the accumulation of real data.

Excuse me for a minute while we delve into the world of geology. St. Paul's Rocks, on the equator, gives up rocks whose approximate age designation is 4.5 Ga. These rocks are prevalent between 30°N and 50°N latitude on the MAR. The same dredge samples have brought up granites and Paleozoic greenstone rocks.

Samples collected by the DSDP corroborate these dates. Leg 37, drillsite 334 core 22 produced an $^{40}Ar/^{39}Ar$ age of 635 +-102 Ma on anomaly 5 (theoretically 8-10 Ma). Onboard scientists pointed out that

" . . . a remarkable phenomenon . . . " of the Mid-Atlantic Ridge in the vicinity of 45°N latitude is that the dredge hauls are consistently 74% silicic rocks in the haul, and some areas consist wholly of silicic rocks. One such area, Bald Mountain (45°13'N latitude, 28°53'W longitude), contains radiometric dates from the granite that range from 1,690 to 1,550 Ma (Proterozoic). Gabbroic rocks drilled at Site 334 on Leg 37 (37°02'N latitude, 34°25'W longitude), yielded a radiometric date of 635 +/- 102 Ma. Similar rocks drilled at Sites 386 and 387, Leg 43 (on the Bermuda Rise) yielded dates of 671, 381, and 323 Ma (Proterozoic-Carboniferous). [The reader is reminded that dredge hauls do not penetrate the surface, so the ocean floor in the region of these hauls is at least that age.]

On the opposite side of the Mid-Atlantic Ridge in the King's Trough area, gently dipping marine strata have yielded Early Ordovician trilobites and graptolites. Paleozoic trilobites and graptolites, in fact, have been sampled at several localities. I wrote to Barrie Rickards in 1999 to verify this. Interestingly, graptolites roamed the oceans from 540 to 320 Ma (Cambrian-Carboniferous). Protolith (original rocks metamorphic rocks form from) ages of several Mid-Atlantic Ridge basalts are 1230 +or - 350 Ma (Proterozoic). At the junction of the rift valley of the Mid-Atlantic Ridge and the Atlantis Fracture Zone (roughly 30°N latitude, 42°W longitude), basalts of the Middle Jurassic age (169 Ma) are found.

Numerous pre-Cretaceous aged localities have been identified at or near the crest of the Mid-Atlantic Ridge from the Vema Fracture Zone (10°41'N latitude, 44°18'W longitude) to the Romanche Fracture Zone (0°12'S latitude, 18°22'W longitude). At the intersection of the Vema Fracture Zone with the median rift valley, Meyerhoff discovered a Late Triassic-Middle Jurassic microfauna in shallow-water oolitic to pelletal limestone outcrops (dates verified by co-authors Meyerhoff, Furre, and Bronniman). The senior authors of this paper would not permit the true age to be mentioned in the text, but limited themselves to the age designation of "Mesozoic." (The genera and species used for age designation had originally been named and described by Bronnimann in 1972.) 140-Ma "Maiolica"-type pelagic limestones have been found in the center of the basin by Enrico Bonatti at the intersection of the

Romanche Fracture Zone and the Mid-Atlantic Ridge. The limestones occur in a deformed region approximately 4-km thick and 200-km long. This feature, lying along the equator, also contains continent-derived quartzitic siltstones of Paleocene and Eocene age.

The equatorial Atlantic region of the Mid-Atlantic Ridge has long been recognized as a region of considerable structural complexity that contains old rocks: continental flood basalts, sericitic mica phyllite, quartzite, carbonaceous shale, brown coal, and many other rocks derived from continental sources, as well as midocean ridge-type and eugeosynclinal alpinotype rocks. Late Cretaceous-Eocene samples are common. West of the central rift is a 200 to 300-km-wide plateau that dips gently westward into the Guyana Basin NE of Brazil; east of the rift valley, the plateau remains elevated and merges into the Sierra Leone Rise west of Africa. The plateau is capped with sedimentary rocks 400 to 1,200-m thick, sections that do not thin away from the midocean ridge crest. The omnipresence in the equatorial Atlantic of so many shallow-water and continental rocks led investigators to propose that a major E-W structural barrier crossed the Atlantic Basin between the Ceara Rise and the Sierra Leone Rise, approximately 8°N latitude.

Precambrian protolith ages of many rocks from the Mid-Atlantic Ridge and associated ridges, from at least Iceland to 30°S latitude, are abundant. These are mainly in the 1.1 to 1.9 billion-year-old (Ga) range.

A 1988-1989 Soviet study found granitic and silicic metamorphic rocks on the western side of the MAR from the equator to 30°S latitude. These rocks have been suggested to have been formed repeatedly throughout the ridge's existence, an interpretation which explain many puzzling geochemical data.

A 2002 conference by the Scientific Council for Earth's Problems was held by the Russian Academy for Sciences. The results, published in the New Concepts in Global Tectonics Newsletter, reinforced many of the data being shown herein. One item stands out among the rest. Boris Vasil'yev made an updated map of the Atlantic Ocean floor age, this time based on rock ages instead of magnetic stripes. This is the first time an ocean basin has been defined in the same geologic terms that are applied to maps based on continental numbers. The map does not look anything like the maps based on magnetics, totally refuting the ideas for this basin put forth by mobilistic reconstructions. Interestingly, the trenches are found to be very young. This study, regardless of others that have shown the same thing, shows the idea of the magnetic measurement, as it is used now, to be totally fallacious. And, this has been known since the first studies done in the 1970s refuted the 1950s ideas.

At this same conference, V.T. Frolov presented facts that clearly demonstrated there was no Paleozoic Ural Ocean, thereby refuting previous work by Sengor (Figure 3). The region has not undergone any significant contraction. Igneous and sedimentary rock studies, along with the geologic formations and styles of folding and faulting, do not confirm extension for hundreds of kilometers. Neither do they confirm subduction, accretion, or tectonic piling. The study does confirm that the Paleozoic Ural Mountains were dominated by vertical tectonism.

As for the DSDP/ODP, the principal investigators have, by direction or otherwise, misrepresented the actual rock ages and types. In a summary article of the history of the DSDP/ODP for the Marine Technology Society Journal in 1998 (page 9), Davies made the quote listed above. The above examples show the extent of that cover-up through dredge and drill data. Examples from the Atlantic basin were used primarily because this basin "opening" is featured so prominently in the Bullard fit, continental drift, and the plate-tectonic hypothesis in general. It is also the most misrepresented. When Davies wrote the article listed above, the age information was available to him from such sources as the *Canadian Journal of Science, DSDP Volumes 37, 43,* and *58, Tectonophysics, Earth and Planetary Science Letters, Science, GSA Memoir 132,* and presumably, all of the *USSR Academy of Science Transactions.*

Who supports the ODP? Why, you do, of course. Think you're getting your money's worth? The Joint Oceanographic Institutions Inc. runs the program. Sounds nice, right? Guess who funds it? You got it, bro, the NSF.

This section does not confirm the 200 Ma maximum age for the ocean floor, which appears to be almost an order of magnitude older based on rock samples. Is Russian science that far off? I think not. I have only given a smattering of the ocean floor data for the Atlantic ocean basin, but this can be said for ALL of the ocean basins. Suffice it to say, we have been delivered a bill-of-goods on this topic. We have been cheated of our tax dollars bigtime. Congratulations, you just added your first ten shares of the Brooklyn Bridge to your portfolio. The ODP is peddling snake oil, folks.

6-2. Midocean Ridges and the Creation of New Seafloor

The first, and most prominent, feature associated with Earth-moving geodynamic events was discovered very early in the history of ocean floor study. The midocean ridge (MOR) was discovered in 1855 when Matthew Fontaine Maury drew the first bathymetric chart of the North Atlantic and called it a "great shoaling" middle ground. Subsequent cruises, such as cable-route surveys by the HMS GORGON and the HMS CYCLOPS, added to the data base. The HMS CHALLENGER Expedition of 1873-1876 found and traced the MOR.

The Navy Hydrographic Office (HYDRO; former name of NAVOCEANO) compiled a bathymetric update from 50°N to 40°S latitude was made in the late 1800s. In 1956 Maurice Ewing and Bruce Heezen found that small earthquakes, found near the ridge axis, outlined a steep-walled linear valley following the crest of the MOR in the Atlantic. Heezen and Marie Tharp had been coming to HYDRO with shopping carts. They would roll out the fathograms across the entire North Atlantic basin. From these Heezen constructed about six profiles, which showed the great "North Atlantic Shoaling Ground," which would become the Mid-Atlantic Ridge. Edward Bullard discovered that the material at a shallow depth beneath the ridge was hotter than the surrounding area (also in 1956). This information was assembled and published as the world's MOR system.

Plate Tectonic Presumption: Bob Dietz and Harry Hess found the rift zone to be geologically active, the stress state theorized to be everywhere in tension along the MOR because of seafloor spreading. The MORs provide the "ridge push" part of the plate tectonic equation. The MOR is not spreading everywhere at once, only where active magma resides.

All rocks there are extremely young and basaltic in nature from a geologic standpoint. The MOR was found to be in tension where the focal mechanisms of earthquakes show normal faulting to build an axial valley with vertical walls. Viscous material, rising in this vertical walled cleft, adds magma to the spreading ocean floor. This is where new oceanic crust is produced, so the basalt is fresh. As the crust moves away from the MOR it cools and subsides so that cracks form and the features become blocky and jumbled. This imprint gives short tectonic spreading fabric (TSF) fractures parallel to the ridge axis. The new ocean floor forming there records the magnetic field at the time of its formation.

The MORs are segmented, being offset by transform faults. Through various tectonic processes, the segments may become offset so that they overlap. The overlapping spreading centers (OSCs) are non-rigid discontinuities. The axial depth deepens by several hundred meters, the geochemistry of the erupted lava changes, and the fine scale structure of the ridge changes. OSCs vary in segmentation length to produce second-, third-, and fourth-order discontinuities. The shorter length, 10-50 km, fourth – order OSC occurs as very small axial summit graben offsets that are barely detectable bathymetrically. TSF produced in this tensional environment fit the description of OSCs.

Segmentation of the MORs is responsible for the pinch-and-swell geometry whose segments average 50-300 km long. The segments are sequentially scaled, beginning with the shortest segment, called the overlapping spreading centers (OSC), or eddy-like structures. The next larger segment, separated by less prominent ridge-transverse depressions, have been called transfer zones, devals, and accommodation zones. The largest have been called transform faults, or ridge-transverse fault zones. These segments actually range between 15 and 500 km in length, and they are a minor player in the overall scheme rather than what they originally referred to by Tuzo Wilson (1965). On-ridge seamounts are formed at the crest of the MOR and sometimes continue to grow as they subside. In the Atlantic on-ridge seamounts exist primarily from the Hayes Fracture Zone to the north. On the basis of morphology and structural evidence, most small seamounts probably originated near ridge-transform intersections and fracture zones and are round, although the shape is thought to be determined by hydraulic resistance to the flow of magma through the edifices. Their life-span is very ephemeral for the most part. On-ridge seamount formation is a function of the magma chamber and supply beneath the rift system. A slow spreading system such as the North Atlantic produces very few seamounts. A medium spreading rate such as the Juan de Fuca Ridge will produce en echelon volcanic chains. A fast-moving system such as the East Pacific Rise produces many more seamounts. Seamount volcanism was more active in the Eocene. Beyond the Eocene the number of seamounts decreases, but the volume increases.

Actuality: We already have seen the age part of this definition to be in error by at least one order of magnitude. Let's go from there. The total lengths of the MOR sections (Figure 2) have been measured at between 75,000-km and 80,000-km: the African Rift system (3840-km; Figure 2-1) and continued through the Red Sea (2000-km), Arabian Sea (1280-km), Carlsberg Ridge (2000-km; Figure 2-2), Mid-Indian Ocean Ridge (4200-km; Figure 2-3), Southeast Indian Ocean Ridge to the MacQuarrie junction (6500-km; Figure 2-4), Pacific-Antarctic Ridge from the MacQuarrie junction to the Eltanin Fracture Zone (3500-km; Figure 2-5), and up the East Pacific Rise (10,000-km; Figure 2-6). While in the Pacific basin the MORs for the Nazca basin (5500-km; Figure 2-7), the Galapagos Ridge (2300-km; Figure 2-8), the Gulf of Alaska (1500-km; Figure 2-9), the Lau-Havre Ridge (2700-km; Figure 2-10), and the Mariana Trough (1400-km; Figure 2-11) conclude that basin's MORs. The Indian Ocean basin MOR from the Bouvet junction to the Mid-Atlantic Ridge is first the Southwest Indian Ocean Ridge (4000-km; Figure 2-12) and the Atlantic-Indian Ridge (2000-km; Figure 2-13). A portion continues to the Scotia basin, that containing 1850-km (Figure 2-14). The mighty Mid-Atlantic Ridge is more-or-less 18,400-km long (Figure 2-15), depending on where you stop in the Arctic Ocean. This figure includes the Reykjanes (Figure 2-16), Kolbeinsey, Mohns (Figure 2-17), and Lomonosov Ridges. Finishing the MORs recognized by most authorities, we will include the Cayman Trough at 1100-km (Figure 2-18) for a grand total of 72,220-km of MORs. This total ignores the recently proposed MOR dividing the India-Australia basin, which would only increase the value by several thousand kilometers.

Seafloor spreading occurs in both directions perpendicular to the ridge, so we have 144,440-km of potential spreading at any one time, depending upon, of course, the magma supply.

Interestingly, right at the start, the term MOR is not true for every ocean because the ridge is sometimes off to one side of the basin, such as the East Pacific Rise.

Because of the veritable plethora of available information, I will only bore you with the northern Mid-Atlantic Ridge, hereafter called simply the MAR. Stacked profiles show the MAR to be shoaler at the crest tapering to low abyssal plains on the flanks. The various highs along the profiles are blocks, ridge-formed seamounts, or sundered plateaus. Superimposed profiles of the MAR at 17°30'N latitude and the front range of the Rocky Mountains at 39°N latitude show such remarkable similarity, that they must have been produced by the same tectonics.

On one of my early surveys, we sailed from Trondheim to look at this section of the ocean. One of my co-workers, Sid "the 300-pound Freckle" Morse, decided he needed a last-minute smooch from his lady friend, who happened to be standing on the dock. As we pulled away, Big Sid jumped over the side and onto the dock, a drop of about 12 feet. Well, you guessed it, he twisted his ankle. Another shipmate, Tom "the Stick" Pettin, was about as big as my arm. We threw a line over to the Freckle and heaved for all we were worth. Remember, the DUTTON is pulling away from the dock, and us two puny guys are hauling this oversized sack of you-know-what back on board for all we are worth. Fjord water is mighty cold. Big Sid laid on his bunk all cruise while we waited on him. He did say that final kiss was worth it all!

Then, to exacerbate the problem, I went down to my stateroom to use the head. There was white powder all over it! Big Sid's roommate had brought a case of crabs on board with him. Very bad beginning.

In a cooperative effort, researchers have surveyed a large portion of the northern MAR. Mohns Ridge in the Greenland and Norwegian Seas extends from 5°W to 7°E longitude. This segment, spanning from 0° to 3°E longitude, has been covered near – totally by SeaBeam surveys (Figure 8). From SW – NE, the axial valley is bounded on the flanks by coalesced seamounts and ridges that range from 1500-2000-m deep.

The 3200-3400-m axial valley is disjointed, segmented, and offset 30° from the overall strike of the MOR. The en echelon valleys are bounded on the ends by large seamounts. The valley floor has many volcanoes and ridges.

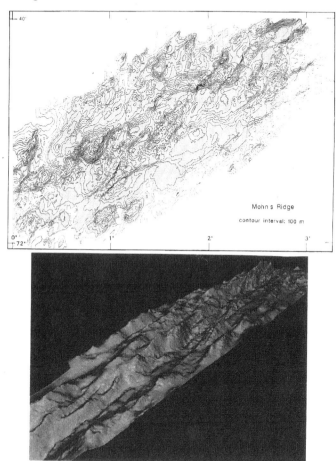

Figure 8. Mohns Ridge in the Arctic Ocean from multibeam sonar surveys at a 100-m contour interval. This total-coverage survey shows a heretofore unsuspected axial valley geomorphology in the form of parallel, but diagonal to the ridge, fracture valleys.

Iceland is between these two segments. Iceland has at least four rift zones, so we may say that, in keeping with the spirit of plate tectonics, it has a diffuse spreading center. Magma wells up in parallel to subparallel fissures, faults, and fractures for the entire length, a region from latitude 55°N to 70°N of the MAR. As we have seen, two of the volcanoes in this spreading center produce up to 75 km^3/yr of lava. The continuous MOR of the MAR at Iceland not only diffuses but it also disappears for over 180-km north of the island to Kolbeinsey Ridge and 330 km NE to Jan Mayen Ridge. This may or may not be a function of the deep volcanic pile created by Iceland itself. The portion of the MAR south of Iceland is the Reykjanes Ridge.

The segment of the MAR that has been used constantly to prove the plate-tectonic theory is adjacent to Iceland. The Iceland Plateau had been thought to have been formed as part of the British-Arctic plateau-basalt province, which had broken up during the initial rifting of the Atlantic Ocean and sunk below sealevel. Because of Iceland's position on the MAR, investigators decided to have Iceland formed on the ridge-crest by volcanic eruptions. In a later explanation, a hotspot was felt to be the creator. Yet later work theorized at least three rifts through Iceland and no extensional tectonics (read seafloor spreading) because the neo-volcanic zone (new rocks) overlay Pliocene stratigraphy (older rocks). The debate papered the press, and in 1992 the magnetic lineations used to "verify" seafloor spreading on the Reykjanes Ridge were refuted.

At this writing the Iceland Plateau seems to have been continental at one time, at least 700 Ma. It was part of a large North Atlantic dam that included Greenland, Iceland, the Faeroes, and Rockall Bank. Spreading is possible only in rift zones, so no spreading occurred in this part of the Atlantic basin. New lavas can hardly overlay older unless there is no spreading.

The MAR between 37° and 35°N encompasses the Project FAMOUS area and the Oceanographer Fracture Zone. From the southern end the rift axis leaves the ridge – transverse fault area at exactly the nodal basin. The first 55-km segment of the MAR on a 015° azimuth is a very pronounced rift axis. Fabric is imprinted on the high 1830-m blocks, or horsts, forming. A noticeable lack of axial volcanoes exists on this segment. The second segment, offset 46-km to the east, is 65-km long and has an equally, if not more so, pronounced rift valley ranging to 3110-3475-m deeps. No volcanoes are present at this contour interval. Here again large, 1800-m relief horsts exist on each side of the rift valley. The third segment is offset 15-km east. At this contour interval, many axial volcanoes are present. While fabric is being imprinted, it is doing so at a lesser rate than further north, presumably because there is less of a magma outpouring for crustal creation. The fourth segment is offset 22 km and is 44 km long. It, too, is rife with axial volcanoes. Apparently the MAR meanders from one ridge-transverse fault to another, or there are more ridge-transverse faults than previously thought.

Photo mosaics of the FAMOUS area of the MAR show the axial valley floor strewn with pillow basalts and broken lava rock debris. This axial rift was found to be a narrow trough approximately 2-km wide, when the first multibeam sonar was collected and presented. The axial rift valley was characterized by low relief, linear ridges on the floor and narrow flat-topped terraces and benches with outward facing antithetic scarps on the walls from 36-37°N.

The portion of the MAR south between the Oceanographer (35°N) and the Hayes (33°N) fracture zones is also segmented. The first 70-km segment north of the Hayes has a clean axial valley which is 2200-m deeper than the surrounding walls. The second 82-km segment is offset to the east by an unnamed 70-km long ridge-transverse fault. The 2000-m deep axial valley is loaded with volcanoes. The third segment is offset 30-km to the east by yet another ridge-transverse fault before continuing to the NE for 93-km and joining with the Oceanographer Fracture Zone. No readily apparent axial volcanoes appear in its 550-m deep axial valley.

The Atlantis Fracture Zone lies to the south of that at 30°N (Figure 9). This 3-D shows that some Fracture/transform axes cross the Mid-Atlantic Ridge, some braid, and some don't cross at all. Considering the fact that this diagram covers about 175,000 km^2, this is no small piece of real estate.

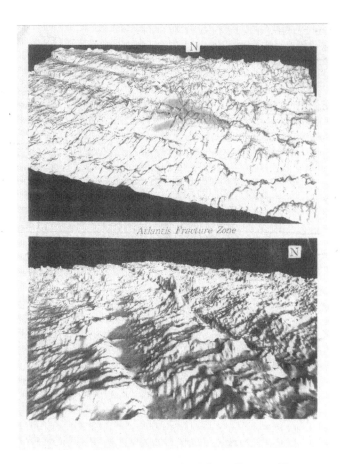

Figure 9. The MAR around the Atlantis Fracture Zone based on multibeam sonar surveys. This diagram shows the ridge segmentation and offsets caused by the braiding of the fractures.

Switching oceans, the MOR for the Pacific Ocean basin, the East Pacific Rise, is on the eastern side of the basin (Figure 2-6). The East Pacific Rise has been extensively surveyed and described by SeaBeam sonar by such programs as RIDGE and RIDGE 2000. The results are easily located on the Internet, so they are not presented here, other than to note their existence. It shows a mixture of flat terrain to overlapping spreading centers. Some of the hummocky terrain is explained by the presence of black smoker chimneys, which may be up to 10-m high. The East Pacific Rise has no clearly defined rift zone.

From a geomorphology standpoint, the East Pacific Rise runs ashore at the southern tip of Baja California and continues north in the form of the mountain ranges to the west of the Rockies (Figure 2). It re-emerges to the north in the form of the Juan de Fuca Ridge, an offset by the San Andreas Fault of some 3000 km, a feature which is called a transform fault. Once again, this is the central, active portion of a fracture zone.

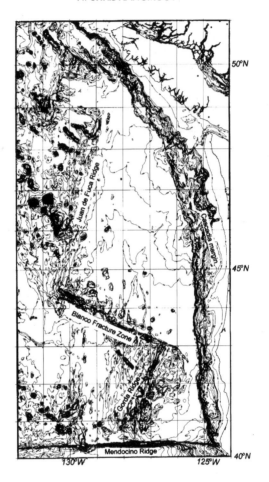

Figure 10. The Juan de Fuca and Gorda "plates" at a 100-fm contour interval on a Mercator projection. This eastern Gulf of Alaska section shows several of the trends, with the seamount chains in Figure 11 on the upper left. Of interest here also is the lack of a trench at the Cascadia margin. The shelf and slope plunge directly down to the plain

The Juan de Fuca Ridge (Figure 10) is a part of the Gulf of Alaska ridge system. The 2380-m deep rift valley is offset west and obliterated by Axial Seamount. Seamount production by the ridge is concentrated on the west, with only the one on the east. Sediment cover in the Gulf of Alaska is very heavy. The Juan de Fuca Ridge eastern flanks are draped with enough sediment to show a bathymetric depth of 2654 m.

Here it becomes necessary to delve into the esoteric world of science for a few moments. Please excuse me, but a point must be made. The other seamounts are parts of ridges that merge to become the San Andreas Fault (Figure 11). From the figure it is obvious that no age sequences occur in any of the three linear seamount chains.

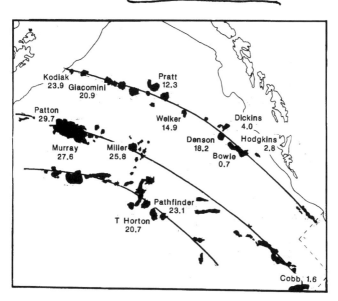

Figure 11. Multibeam sonar-based tectonic figure of the Gulf of Alaska seamount chains with age data from various sources. This diagram amply demonstrates the ESE-flowing trends and the intra-chain age differences.

On another DUTTON cruise, this time in the late 1970s, I was on my fifth survey in the Gulf of Alaska. It was in January, which was a helluva time to be up there anyhow. Three low pressure cells converged on us. We could not run because we had nowhere to go. In what I would have to call the stupidest thing I have ever seen, much less heard tell of, we made a 180° turn in 54-foot seas with 95-knot winds. Well, when we got through the fanny-spanking pitching and the awful rolls, we did manage to hove to for a few days until the storm died. We limped into Oakland and found out the ship had 24 cracks in the hull forward of the super-structure. We were lucky to be alive, some said. I felt lucky!

The tectonic interpretation of the function of the MOR has been in a state of flux. Sidescan sonar imagery has not only given high-resolution coverage but also much continuous coverage over thousands of square kilometers. Observations based on actual data include, aside from the transform faults every 35 km or so in certain regions, the occurrence of non-transform ridge axis discontinuities such as overlapping spreading centers (OCSs) and en echelon faults, fissures, and fractures that parallel the spreading axis. This phenomenon reflects Stokes' Law; essentially Newton's Second Law of Motion. Magma flow is indicated to go along-strike on the ridge instead of away from the ridge. Poiseuille flow patterns are flow-parallel shears between different velocities, such as in lava tubes, glaciers, and the MOR, and the parallel cracks, in turn, are the geomorphic expression of Poiseuille flow. The TSF has by now become the ridge-parallel fault-fissure-fracture (FFF) fabric. The Reykjanes Ridge, the FAMOUS area, the Juan de Fuca Ridge, and the East Pacific Rise laminar flow features are offset at the recognized transform faults, with several of the transform faults overprinted. The fissures, faults, and fractures continue unbroken in a N – S direction.

The MORs are everywhere characterized by the FFF systems, even out onto the flanks. These are stretching lineations, and they continue onto the continents in many cases, such as the East Pacific Rise. This particular FFF bathymetry is characteristic of all tectonic belts, so that description includes not only MORs but also aseismic ridges, linear island and seamount chains, oceanic active volcano arcs, and foldbelts. Essentially, the magma flows along strike, not perpendicular to it and out into the world. Oh, much leaks out to overprint the 1.5 Ga rock already in place, don't misunderstand that. But, not enough, and certainly not strong enough, force is exerted by the leakage to push the opposing plates apart. No hydraulic press effect can exist because the magma chamber is not a closed system. Therefore, ridge push as a driving force of plate tectonics is a figment of an overly active imagination.

If you can fast-forward to Figure 13, you will notice that on-ridge formed seamounts are a rarity in the North Atlantic, so rare that almost none exist in nature. On-ridge seamounts actually are the famous black smokers, worm tubes, etc. that are in the popular press. According to the total coverage surveys that I looked at, and that was all of the Navy's multibeam data over the years until 1998, I found no on-ridge seamounts with the exception of Iceland and Ascension Island in the South Atlantic. One must remember, though, that most people are constrained by definition to describe seamounts as "elevations rising more than 1000-m and of limited extent across the summit;" 50-m hills do not count as seamounts.

From this section we will take along-axis magma flow and the existence of a world-wide web of interconnecting magma channels. You just purchased another 15 shares in the Bridge. Do not pass GO . . .

6-3. Convergent Margins

On one of my first cruises, early in 1967, I was on the MICHELSON, or Mickey Maru as we called it. We surveyed a region called the Nero Deep, which is somewhere in the NW Pacific trench region. This was soon after Don Walsh had made his famous dive down into the Mariana Trench in the bathyscape Trieste. On my last cruise, as we were trying to outrun a super-typhoon going from the South China Sea to inport in Yokohama, we stopped for an ocean station in the Ryukyu Trench. On my watch, as a going-away present, the ship took a water sample from 6000 meters. I was told this was the deepest water sample ever taken by a NAVOCEANO ship. They gave me a bottle, and, sure enough, it tastes like salt water!

Plate tectonic Presumption: "Slab pull" at the active margin subduction zones, one of the basic tenets of plate tectonics, is a second driving force of that hypothesis. Several types of subduction zones exist. (1) In a collision between oceanic lithosphere, one subducts and the other overrides. The overriding lithosphere does so because it is warmer due to the presence of the active margin channel. The subducting plate is called "seaward," and the overriding plate is called "landward." (2) In a collision between a continental plate and an oceanic plate, the oceanic plate always subducts because the oceanic lithosphere is thinner. (3) In a collision between continental and oceanic plates where there is also oceanic crust present seaward of the continent, the landward oceanic crust can subduct the seaward oceanic crust first. Finally (4), two continental plates meet in a continental collision. This process gives a suture zone, or mobile belt, which is a closed subduction zone. A large amount of horizontal compression and overthrusting always exists in these settings. All of the intervening oceanic crust is consumed, followed by the abandonment of the subduction zone. The shallow sea in between is filled with sediments, and the mountains begin to rise. These sediment deposits are folded, and igneous rocks are intruded into them.

A passive margin (the shelf, slope, and rise from the plain) is non-volcanic or orogenic. The sediments in the shelf area show abundant evidence of shallow-water deposition and, characteristically, prograde seaward; that is, they get thicker. In the plate tectonic hypothesis, continental rifting precedes passive margin subsidence. It begins with graben formation followed by volcanic activity, such as basaltic dike and sill intrusion. A passive margin can become an active margin when subduction begins along it or because an ocean between it and a land, beneath which the sea floor is subducting, closes.

Actuality: In this discussion, the trench is the primary feature of the active convergent margin, with other features acting or interacting around the trenches. They are: (1) the Pacific Ocean basin has the Kuril-Kamchatka (2200-km), Japan (800-km), Izu-Bonin (800-km), Mariana (2550-km), Ryukyu-Philippine (1400-km), New Hebrides (1900-km), Vityaz (2500-km), Tonga (1400-km), Kermadec (1500-km), Peru-Chile (5900-km), Middle America (2800-km), Cascadia (1600-km), and Aleutian (3700-km), (2) the Atlantic Ocean Basin has the Puerto Rican (1550-km) and the South Sandwich (1450-km), and (3) the Indian has the Andaman-Nicobar-Java-Timor (4500-km) for a total distance of 36,550-km (Figure 3).

The *raison d'etre* of the trench is a function in and of the active, convergent margin. The convergent margin is comprised of the seaward lithosphere, the trench which has inner (landward) and outer (seaward) trench walls and an axial low, and the landward lithosphere. Inner trench walls can have terraces, perched basins, obducted fragments or ophiolites, re-entrants, and accretionary sediment prisms. The outer trench wall is usually characterized by horst and graben structures from normal faulting and occasional aseismic bathymetric irregularities, such as seamounts. The landward lithosphere contains the forearc and is overprinted by the active volcano arc in most cases. The width of this particular zone varies.

The trenches are usually underlain by earthquake activity, and that is discussed later. By and large, cold, oceanic (seaward) lithosphere comes in contact with warmer (landward) lithosphere that does not have the same pedigree. Because the seaward lithosphere is cold and thin, it descends along crustal cracks to a depth of about 100 km and appears to bend toward the landward crust.

The convergence process may be temporarily terminated by the introduction of major bathymetric highs. The collision of large, aseismic, bathymetric highs will permanently alter the trench's geomorphology.

One rather prominent feature is associated with active margins, and that is the cusp. While only a few exist in nature, numbering probably less than fifteen worldwide, they are large enough to elicit comment. First noted to be where large aseismic oceanic ridges were residing on the seaward lithosphere, they were suggested to be the effect of subducting these ridges, ridges which appear to be resisting subduction in the plate – tectonics scheme of things. Throughout the rest of this chapter, many of these cusps will be discussed. These include the two continental cusps surrounding the Indian sub-continent as well as those lying in the northwest Pacific Ocean basin. While the final disposition of the aseismic ridges in their collective dotages may be the same; i.e., they will disappear as an entity, the emplacement mechanism will change appreciably.

Having discussed the different features that comprise an active convergent margin, we make this observation from a geomorphologic standpoint. Nearly all of the subduction zones are in the Pacific basin and face the east. The mechanism for this phenomenon was postulated to be an eastward flowing mainstream existing in the upper mantle that was connected to the convection cells beneath the midocean ridges.

We are dealing with bathymetry, and the bathymetry will give us all the clues we need about subduction zones. The Cascadia "Trench/ subduction zone" is noticeably absent. There is a reason for that. In the bathymetry (Figure 9), no trench exists. In the earthquake data, only scattered shallow events appear. No trench; no earthquakes; no subduction zone. Scratch about 1600-km from the total. The same holds true for the Vityaz "Trench." A series of profiles made across that region revealed that the seaward ocean floor is deeper than the proposed trench (5400 m vs. 5200 m). So, we can remove that "subduction zone" distances of 2500-km from the equation, leaving us with 32,450-km of available subduction zones.

The trenches have always been shown as being continuous from start to finish and generally the same depth for each trench respectively. Ted Ranneft, an exploration grologist who spent many years working in the SE Asia region, first theorized that trenches may actually be discrete segments, and that these segments are at different azimuths and depths. He showed this for the Andaman-Nicobar, the Java, Timor, and Philippine trenches in a 1979 paper. Trench segmentation may show many variables: (1) different types of volcanism may exist across these boundaries, (2) different types of volcanic products may also occur, (3) the separate segments may move independently of each other, and (4) extremely virulent or unusual volcanoes may exist in these gaps. By using the variable trench depths, this paradigm may show segmentation where none exists bathymetrically.

A sampling of the variable depths includes: The Aleutian Trench ranges from 5800-6900-m, Kuril, 7000 m; Japan, 9000 m; Kashima guyot to Delaine Seamount (Izu), 8600 m; Delaine Seamount to Uyeda Ridge (Bonin), 9000 m; Fryer Guyot to Hussong Seamount (Mariana), 7000 m; Hussong to delCano Seamount (MAriana), 8200 m; south of Serrao (southern Mariana/Challenger Deep), 9600 m; Manus, 6000 m; Kilnailau, 4500 m; New Britain, 8000 m; Tonga, 6500 m; Kermadec, 7500 m; north Peru, 7000 m; and Chile, 5300 m.

Where the trenches are segmented, obduction has been the historical explanation. This is also the site of most of the seismic activity and the clustered volcanoes. Where Ranneft demonstrated his ideas, presumably based on land-based transverse faults and oil-field data, multibeam sonar bathymetry has substantially verified those ideas.

Tectonic events at active margins are explained by Ranneft as oblique faulting all over the Pacific Basin. He says that the "hinges" may be related to structural, morphological, or movement aspects of the underthrusting lithosphere in the Aleutian Trench region. The fold axes, which occur as a first-order deformation, are 030°, 060°, and 080° for the entire length. The 060° axis is perpendicular to the convergence angle of oceanic lithosphere. Thrust faults ranging from 055° to 095° can be mapped upslope from the anticlinal folds as the second-order compressional deformation. Further up the inner trench wall oblique faults occur. West of 160°W they trend 280°. East of 160°W they trend 020°. In the Aleutian "trench" a 60 Ma old transverse fault transects the two which passes through an aseismic zone coincidental to Unimak Island. No aseismic gap exists at 160°W, although the location of the old transverse fault is fortuitous in its proximity to the clustered volcanoes across the trench from Derickson Seamount.

Bathymetrically, the idea of a smoothly descending slab seems a little farfetched. We have one other tool to verify the existence of subduction zones, though, that of earthquakes. As we saw in the introductory material, earthquakes were used to develop the idea of the smoothly descending slab in the first place.

The trench will be underlain by a line of earthquakes, which get progressively deeper the further one moves landward of the trench. In fact, the latest from the in-crowd is that slab-pull is the primary driving force of plate tectonics. This will be shown to be the largest of the sticks of globaloney. Collision margins are defined by earthquakes, so that the earthquakes are an important adjunct to the study of convergent margins. Most importantly, though, earthquakes are a natural hazard to reasonable living conditions.

Earthquake data show activity at depth in the upper mantle, which occurs midplate under the midocean ridges, along fracture zones, under volcanoes, etc. More often the activity is located at the active margins. The earthquake regime is defined as: shallow earthquakes down to 70 km, intermediate earthquakes between 70 and 300 km, and deep earthquakes deeper than 300 km. The frequency decreases logarithmically down to 300 km, then increases again to 550 to 600 km. No earthquakes are known beyond 660 to 700 km.

So, in a recap of an older study made in the 1990s, I plotted all of the earthquakes in the NGDC data base, color coded at 50-km depth increments. A closer look at the data for the convergent margins is now presented for your edification and amusement. The earthquake problem is addressed in relation to the location of the deep earthquakes. The National Geophysical Data Center (NGDC) is the repository for worldwide earthquake information. A plot all of the earthquakes, over 180,000 events, from the 1994 data base showed that 88% lie above 150 km. All of those are related to "magma" (I have put the quotation mark because the conventional understanding of magma is a hot liquid, but as we will see the rising material is mostly cold and solid, i.e., solid plasma) movement, either outlining horizontal channels or stacked vertically in the form of rising "magma" to produce volcanoes or fractures.

Table I. Summary earthquake data for all convergent margins.

(total number of earthquakes in study from NGDC earthquake 1990 CD-ROM is 182,597)

Earthquake depths (kms)	Quantity	% of total	
0-49	121,557	66.6	
50-99	25,633	14.0	
100-149	14,481	7.9	
150-199	6941	3.8	
200-249	4008	2.2	
250-299	1357	0.7	
300-349	801	0.4	supposed increase at
350-399	856	0.5	eclogite phase change
400-449	905	0.6	
450-499	871	0.5	
500-549	1569	0.8	increase at spinel
550-599	2060	1.1	phase change
600-649	1341	0.7	
650-699	212	0.1	
700+	5	-	
unknown	5300+		

The deeper earthquakes, numbering 4.3% of all events, seem to reappear at about 350 km, go down to 650-km, and are related to another form of tectonism. What deeper earthquakes existed in the data base were aligned with aseismic, buoyant highs on the seaward sides of the trenches, and that this occurs in only nine spots worldwide. The groups (Figure 3) are located (1) at the juncture of the Mediterranean-Indonesian belt with the Tonga Trench south to New Zealand, (2) a few north of Sicily, (3) between 6°S and 11°S under South America, (4) between 19°S and 29°S, also under South America (5) behind Kamchatka, (6) from 26°N to 36°N behind the Bonin arc, (7) a few north of the New Britain Trench, (8) a line behind the southern Philippine trench, and (9) a line north of Java and Timor. All of these are on the northern portions of their respective active margins, and they appear to be segmented. Previous deep earthquake studies do not apply for the rest of the real world. In actuality, if the SW Pacific convergent margins were removed from the table, negligible earthquake activity below 250 km would exist globally. This means that only one active margin fits the definition of subduction zones from a seismologist's standpoint.

What this means is that the subducting plate, melting back into the mantle at 650 km depth, is a myth. But, this is not recent news. Benioff himself actually showed this to be the case in the 1950s, when he plotted several of his deep thrust faults. He showed that the upper earthquakes, on a shallow angle, were totally disjointed from the deeper earthquakes, which were on a steeper angle. It seems as though a mere flourish of the cartographic pen joined the two mutually exclusive groups into a continuously descending slab. This idea had been proliferated since the 1950s unabated, even though the earthquake data has been universally accessible. What this does is to refute the concept of subduction; no continuously descending plate exists.

Now Dong Choi, an exploration geologist consultant in Australia, has made hypocenter plots of the earthquakes at certain margins. Where I had shown a lack of intermediate depth earthquake events, he has gone one step further. The deep earthquake events all lie in a region of supposed deep fault zones, they are generally stacked, and they are delineated on the surface by tectonic belts. For the western Pacific this means trenches and active arcs. For the eastern Pacific this means uplifted features such as the Andes Mountains.

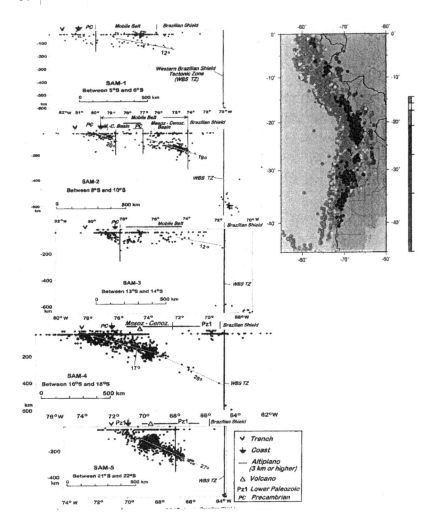

Figure 12. Earthquake hypocenter plots off west-central South America. These plots, which Dong Choi used to expand upon Benioff's original plots done 50 years before, show the discontinuous descending "slab" and the two different angles outlined by the earthquakes. The earthquakes are from 1996-2002. (used by permission of Dong Choi)

Choi's most recent study includes the Peru and Chile trenches off the west coast of South America (Figure 12). Where I had color-coded the epicenters, Choi actually gave depth to his study. In a series of six lines across the trenches, he found that almost all of the earthquake activity was above 200-km. It also shows a variability of descent angle, intermingled of 12°, 19°, 12°, 17°, 27°, and 20°. This is a really crooked plate! While the epicenter portion of Figure 12 shows deep earthquakes, the size of the dots precludes a real understanding of events there. In all six of his lines, no events are noted between 200 and 600-km. In other words, no intermediate events occurred. No continuously descending slab exists, and all of the deep quakes are clustered below the Western Brazilian Tectonic Zone. Friends, we're talking about a region from 5°S to around 27°S in two of the nine available spots worldwide for deep earthquakes to appear. This is not good for the home team.

Choi concludes that deep earthquakes are related to deep tectonic zones which are recognizable on land and that the deep earthquakes occur in subsiding areas. They are directly responsible for the subsidence of the upper mantle and crust along the major deep fault systems, such as the western Pacific and the Peru and Chaco-Panana Basins off South America. He has found that the seismic focal plane leans oceanward, which indicates that the Benioff zone is a reverse thrust fault system.

A first-order estimate from all of the above leads to the conclusion that the inferred subduction or engulfment, process – I say inferred because nobody was able to prove experimentally that a thermally driven penetration of one solid into another solid is possible, could take place only in the lithosphere; that is, the upper 100 km of crust, the supposedly more rigid – according to the PT people – part of the earth's interior. In other words a mess of *ad-hoc* interpretations and logical contradictions. All of the convergent margins have numerous earthquakes in the upper 100 km. As a matter of fact, these shallow earthquakes also pervade the mid-ocean basin lithosphere, too. Anyhow, below that, no subduction exists. At least for the Pacific Basin at the Aleutian "trench," the Kuril and Japan trenches, and the Middle American, Peru, and Chile trenches, seismic profiles that show the same Layer IV lithosphere on either side of the Benioff zone, and that continues onto Precambrian continental crust (Figure 22). Friends, that includes most of the trench linear distance, like about 12,600-km of the remaining 32,450-km.

The notation is made that the shallow earthquakes underlie major ocean floor highs, such as the island arcs. Anfiloff calls this "basement ridge tectonics," which means that the ridges are surficial expression of compression caused by Earth contraction. Because the shallow earthquakes lie atop the deeper ones, they are necessarily younger. This means that the island arcs, trenches, and backarc basins are also young features. Also, large magma reserves lie between 100-300 km below the active arcs, and I will discuss this later.

For the reasons listed above, the entire "subduction suite" is closely associated, and most if not all of the tectonic activity lies oceanward of the stable continental cratons.

In 1998 our bunch of malcontents had a symposium in Tsukuba, Japan. A group of field hands called themselves the "New Concepts in Global Tectonics" working group, and they would meet every couple of years to discuss any new findings and to try to synthesize some kind of meaningful explanation about Earth geodynamics. These are people working in the field trying to use the ideas of those in the ivory towers, and the ivory ideas did not work in most instances. When you're getting paid to perform a useful task, such as find coal or oil deposits, theories don't cut it, and that's what these folks were all about. Anyhow, we took a field trip around the Boso Peninsula. The Japanese hosts did an absolutely outstanding job of everything associated with this symposium, by the way. We were supposed to see the results of subduction, observing the lithology, faults and folds of Pliocene and Miocene formations, and abrasion platforms resulting from regional destructive earthquakes. The excursion leaders were convinced that the stratigraphic reconstruction could have been the result of vertical tectonism just as easily. They also said that subduction under Japan was a physical impossibility because the fault planes of the shallow and intermediate earthquakes deduced from the P-wave patterns were related to block-like deformation. Everyone agreed, which is anomalous in itself!

Recently the *ad hoc* committee consisting of several seismologists and tectonicists have derived a new "proof." Their conclusion is that faulting occurs in the descending crust (lithosphere and mantle material), and this accounts for the shallow earthquakes. As the hydrated oceanic crust sinks it is slowly heated, it becomes dehydrated and becomes "fluid-assisted faulted." The slab's interior remains cold so that the inner olivine in the subducting slab cannot transform into spinel. At 300 km the outer slab temperature increases and enough dewatering takes place to allow anti-crack faulting as the spinel phase changes begin. This process creates intermediate earthquakes. As the slab descends deeper, the inner olivine follows suit, and earthquakes again increase to about 600 km. At that point all earthquake activity ceases.

The astute reader will notice that none of the proposed subduction zones in the Mediterranean region are listed. The Aegean and Adriatic Sea portions have been studied extensively. The earthquake epicenters plot in a time-sequenced, clockwise, flowing circle. No circular subduction zones exist in the classic definition.

To show the depth of the PT hypocrisy, the program for the 2004 International Geological Congress includes four session directly using "subduction" in the title of the symposia. This does not include the field trips to look at the results of subduction related processes and ophiolites.

The trenches are segmented; they do not have the same depths; they only comprise about 20,000-km linear distance. What the bathymetry started, geophysics finished. Once again, it would appear as though we have bought a few more shares in the Brooklyn Bridge, because subduction, an integral one-third of the plate-tectonic formula, does not exist as it has been defined. No subduction, no slab pull. Is nothing sacred!?

6-4. India and the Tethys Sea(s): Polygamy Was Not a Part of Okeanos' Program

Then we have the collision margins. These generally lie from the Mediterranean through the Arabian peninsula, and across the Himalayas. Naturally, all of this earlier movement should have created the sutures found in the present-day continental land mass. One of plate-tectonics' problems is that the model does not recognize that ALL tectonic belts have a common origin, and that compression, strike-slip, and extension all take place at the same time in the same province. Were the India/Asia collision to be a fact, a continent/continent convergence zone would exist, and that is called the Indus – Yarlung-Zangbo Suture Zone by the plate tectonicists.

Plate Tectonic Presumption: As we saw in the Introduction, the Tethys Sea formed between Gondwanaland on the south and Laurasia on the north (Figure 4) according to Seuss. We also saw Sengor giving us two

Tethys Seas (Figure 5). The migration of India and part of the Arabian peninsula across the Tethys Sea basin from south of Africa to Eurasia is an integral part of all plate and expansion models, an event which began 65 Ma. In its wake, the Indian Ocean was forming.

Actuality: This is almost like taking candy from a baby. How is it that the august body of academicians has never bothered to use any of this data, data that is readily available to even the most myopic of students of earth geodynamics? Let us not rest here; let us proceed further into the morass of misinformation we have been spoon-fed for these past 35 years:

The first instance is the Zagros Crush Zone between the Arabian peninsula and Iran. This stepped feature begins in the Persian Gulf and climbs its way, ever northward and progressively higher, to build the Zagros Mountains. No trench exists in this region, merely the stepped signs of compression. The next in the transition from the Mediterranean to the Himalayas is the Makran convergent margin. It begins in the Gulf of Oman and climbs its way up to the Dasht-e-Lut, which is an offshoot of the Zagros Mountains.

Plate Tectonic Presumption: Eduard Suess, a Swiss geologist, introduced the concept of a Tethyan Sea in 1893, and Celal Sengor carried his idea one step further in 1984 by giving us two Tethys Seas (Figure 5). By assuming that Okeanos was a polygamist, that is, that not one but two Tethyan Seas existed, Sengor began his evolution of the Indian region with Late Paleozoic rifting along the northern margin of the supercontinent called "Gondwanaland." This was ancestral to the Alpine-Himalayan belt. No ocean or seaway is supposed to have existed in this region before the Carboniferous time. The rifting caused the separation of a string of continental masses, called the Cimmeride continent. The northern drift of the Cimmeride, a result of spreading in the neo-Tethys Sea, caused the closure of the Paleo-Tethys Sea further north. The Paleo-Tethys was a vast, eastward widening embayment between Laurasia and Gondwanaland. The northern drift not only closed the Paleo-Theyths but it also created the Triassic Cimmeride orogenic belt. This concept of Sengor's was supposed to answer a major problem in plate-tectonic reconstructions of that time by placing large portions of Mongolia and Central Asia within the Tethys.

An updated variant of the Neo-Tethys model, which seems to have found a broad acceptance by the cognoscenti, envisages the origin of this ocean, during the Permian, along the southern margin of a basin that had existed since the Precambrian on the northern margin of Gondwanaland; that is, along the northern boundary of the Indian subcontinent. Remember, India was supposed to be way down south joined to Australia, Antarctica, Africa, and South America. While this model had evolved into a radical shift to that of the original premise, proponents of this model neither support nor refute Sengor's model, nor do they specify the relationship of this Precambrian-Permian basin with his Late Paleozoic-Mesozoic Neo-Tethys or Paleo-Tethys Seas. It also leaves unaddressed the underlying problem that Sengor's original model was supposed to have solved.

However (always the "however"), neither of the two models addresses the stratigraphic and tectonic records of one of the best-known segments of the Neo – Tethys-the Himalayan belt – where the southern passive margin of this ocean preserves a concordant stratigraphy from the Late Archean to the Eocene, a rock sequence recoding events for the past 2.51 billion years. And, this concordant record has been known for at least the past 20 years, in English, and published ubiquitously.

Actuality: Finally, we get to the Himalayas. Ismail Bhat, 25-year veteran of the Wadia Institute of Himalayan Geology, collected rocks all over the Himalayan region and analyzed their composition and age. Here is what he has found:

Following the initiation of basin formation in the Himalayan region during the Late Archean rifting event and associated mafic volcanism (Rampur flood basalts), the sedimentation history reveals a progressive northward shift in the fall line and depocenter that was determined by a similar shift in the locus of the four repetitive rift – related mantle-derived magmatic phases over the time until the lithosphere rupture was completed during the Late Cretaceous. Unlike other rifting phases, the early Paleozoic rifting was not associated with mafic magmatism. Instead, it caused large-scale crustal anatexis which produced a long, linear belt of leucocratic granites in the Higher Himalayas. All the rifting events are marked by the uplift of the basin, the basin which eventually became the Himalayan Mountains we see today. While the Late Archean and Early Paleozoic rifting phases resulted in the establishment of stable rift shoulders, such as the Shield and Lesser Himalayas respectively, the later events failed to do so. This is typically seen in the collapse of the stable rift shoulder following the Late Paleozoic rifting. The difference in post-rift sedimentation could be attributed to the failure in the establishment of a coupling between the surface processes and tectonics.

The Middle Proterozoic saw the Bafliaz and possibly the Abor volcanics, the Late Paleozoic-Early Mesozoic the formation of the Punjal Traps, and the Early Cretaceous was witness to the creation of the Indus-Yarling-Zangbo suture zone ophiolites. These last features, the IYZSZ ophiolites, are attributed to the Tethys Sea depositional regime. However, the lithographic section of this ophiolite, with an interbedded succession of sedimentary and mafic/ultramafic litho-units and a sill complex, dispels any myths concerning the Tethys as an open ocean. It has neither an interbedded succession of sedimentary and mafic/ultramafic litho-units nor a sill complex, both of which are part and parcel of the plate tectonic definition of an ocean floor ophiolite complex. The hard-core field data instead conform to a high-density magma upwelling in a basin having undergone severe extension and rapid sedimentation.

And, of course we are all familiar with the high-mountain marine fossils. Of the 26 km thick concordant sediment record, pre-Carboniferous marine fossils are found in the first 16 km. This sediment pile lies just south of the IYZSZ, which happens to coincide with the proposed subduction zone of the Neo-Tehys Sea. Sengor has no room for a seaway during this time, yet the depositional record contains over 600 Ma of these fossils. Late Proterozoic-Cambrian marine fossils are overlain by huge stromatolite – bearing carbonate deposits all along the Lesser Himalayas. Additionally, the oldest Lesser Himalayan lithologic units are co-eval to those of the Aravalli-Vindhyan region of the Indian shield.

In summary, the Himalayan stratigraphic record following the Late Archean basin formation suggests that sedimentation occurred from the shield region until the Early Cambrian all the way up to the Higher Himalayas. Marine conditions came to an end by the Early Paleozoic for both the shield and Lesser Himalayan regions. It continued to the north where the basin underwent differential uplift and erosion with widespread granite magmatism, accompanied by sedimentation. This produced the Higher Himalayan Cambrian-Ordovician granites. From that time forward not any of the Himalayan region was under water, so India did not "dock" where it is 15 Ma. There was no Neo-Tethys; there was no Gondwanaland.

Sutures (mobile belts), or old collision zones, have been theorized where belts of ultramafic rocks exist on continental cratons. Increasingly, detailed field work has shown that these are not zones of collision at all. Many of the proposed sutures in Asia are now recognized as old tensile, rifting, tectonic environments where mafic and ultramafic rocks have risen to the surface along fault systems. This means that the theories about Asia being a collage of micro-continents is in serious doubt, and that the Sengor model (Figure 5) is globaloney.

So, once again we have a glossing over of the facts, the stratigraphic sequences which were known by all the literate world for decades. The rapid acceptance of this preposterous idea is all the more disturbing, considering field geologists and tectonicists did not question the model. The tectonics notwithstanding, the complete disregard for the rock record for the past 2.51 billion years (beginning with the Berinag group) is deplorable, but, alas, a fact of life in these plate-tectonic times. Fortunately, the thinking person is able to read the facts and digest them as applicable; in other words, a word to the wise is sufficient. Caveat emptor.

Removing this from the possible dump site of the new ocean floor material leaves us with a figure of 32,450-km of take-up for a possible 178,000-km of linear distance at the production end. The math doesn't work. This has already become a moot point, though, as we have clearly demonstrated that the new ocean floor lies atop old ocean floor. Additionally, India has always been just where it is today.

But, what about seafloor spreading? The very concept gets the older lithosphere out of the way by horizontal movement, doesn't it?

6-5. Seafloor Spreading, Megatrends, and Intersections

Walter Smith and Hubert Staudigel got in touch with me during the 1980s. They had a grant to look at the ocean floor between Hawaii and Guam in preparation for a cable laying. A feature I had contoured, named, and written about, Jaybee Guyot, lay along that route. They called me and asked if I had any detailed info to help them out, to which I responded favorably. Later, they called to find out why the feature was named Jaybee. I told them that I had named it after my wife. They asked me if they could do anything for me, so I inquired as to the nature of their investigation. When they said that they were going to collect some rocks, I asked them to send me some so that I could have a necklace made for my wife. They thought this was cool. About a month later I got a Marisat message from Walter stating "Operation Necklace complete. Give my regards to your lady." Now we were getting really cool. A month or so later I got a package from Lamont. Sure enough, the guys had sent me some rocks. I had that necklace made, and JB wore it to a couple of AGUs to show it off. At that time, I figure she was the only person in the entire world to not only have a feature named after her but also to have a piece of jewelry made from same.

As an aside, this feature name was not accepted by the USBGN for those of you who are keeping score.

Plate Tectonic Presumption: Fracture zones, locked and in place rather than active, show the direction of seafloor spreading at the time they were imprinted, or formed.

Actuality: We need to observe what the bathymetry/topography is telling us because we can physically lay eyes on the ocean floor through the newer sonars. Fracture zones were defined as extensive linear features of unusually irregular topography of ocean floor characterized by more than one kind of feature such as seamounts, asymmetrical ridges, troughs, or escarpments having age and regional depth contrasts across the younger end of fracture. This definition had nothing to do with the tectonics except that, through common usage, they were felt to be either active or fossil plate boundaries. Within an active plate boundary strike slip faults predominate. The topography was thought to be the result of differential vertical adjustment with possible extension and compression components. Theory has the direction of seafloor spreading shown by the fracture zones, a concept which allows one to infer an ocean basin's history. If the lithosphere is in motion, the direction is the same on both sides of the fracture.

During the early 1980s, as I was getting cranked up on my guyot project, I kept running into these long lineaments. The Meyerhoffs had already made the observation that, for the Pacific basin, the fracture zones could not possibly show the direction of seafloor spreading (Figure 6); the fracture zones defined a fanning pattern across the basin, all converging in the west-central basin! No comment from the peanut gallery? I decided to try to update that diagram based on the DBDB-5 and what I had already published from our data bases at OSP. The result was astounding; disregarding the proposed elbows at 43 Ma, and disregarding the Meyerhoff's work, led to a checkerboard pattern for the entire Pacific basin. And, it showed the linear seamount chains to be extensions of the fractures. This was of course a physical impossibility, and I spanked my own hand.

By the mid-1980s NAVOCEANO had completed the surveying and charting for the northern Atlantic Ocean basin. The charts were photographically reduced and pasted on a grid, producing a "superchart." From this, I compiled and published a seamount and fracture zone locator diagram (Figure 13). Fracture zone trends in the Atlantic Ocean basin had been thought to lie mostly WNW – ESE in parallel-to-sub-parallel straight lines. Every one of the linear seamount chains was at the distal ends of the fracture valleys, which extended the lineaments. Also, the fracture zones did not lie at discrete intervals. They were not parallel-to-sub-parallel. Fracture zones are not straight. They splay, merge, and stop and start indiscriminately within the constraints imposed by the lithosphere and stress regime. Fracture zones became "fracture zone swarms."

Fracture zones were defined as extensions of the transform faults on the midocean ridge spreading centers. This idea is not even remotely accurate; at least half of the fracture zones in the Atlantic basin are not even associated with transforms. They do not cross the midocean ridge. Instead, they are overprinted by it. This means that the fracture formed before the midocean ridge, because the fractures continue on both sides of the ridge. The fracture zones in the Pacific basin that approach North America are not associated with any ridges, yet the magnetic stripes show them to be still forming.

Figure 13. Multibeam sonar-based tectonic diagram of the North Atlantic ocean basin showing all of the fracture valleys and the seamount locations on a Mercator projection. This is the diagram that some say is not accurate because the positions have been moved. At this scale, they have not been moved even a skochie bit; they are exact. The fractures show later piling on penalty imposed on the Mid-Atlantic Ridge magma channel, which apparently formed after the fractures.

Fracture zones, as zones of weakness in Earth's crust, are associated with mid-plate seismic events, suggesting that their provenance may be related to the buckling and fracturing of the crust caused by the release of mid-plate stress. At any point of weakness at any time magma leakage through the fracture zone can produce seamounts or islands. While the original thoughts had the seamount chains to be age sequential; that is, to progress from younger to older, that has since been proven not to be the case. Seamount chains are the result of normal tectonic activity, and one finds all-aged seamounts and islands intermingled. While this discovery precluded the idea of hotspot generated seamount chains, the data have received very little attention. The presence of a fracture zone can then be determined by the presence of linear seamount chains along with the familiar strike-slip ridge-valley configuration shown above.

For many years now the fracture zones have been known to braid, meander, and splay. For that reason alone, they cannot show the direction of plate motion. Using a highly refined bathymetric surveying system, fracture zones are known to exist that are orthogonal to the fracture zones used in the early hypothesis. The San Andreas Fault is a perfect example. The Mendocino and Murray fracture zones have been proposed to abut the San Andreas Fault at a 90° angle. This fact alone should have raised a red flag. Later research has the Mendocino and Murray fracture zones coming ashore, the Mendocino going all the way to Yellowstone National Park in Wyoming. In fact, most of the fractures do not stop at the continental margins; they continue on through the continents to emerge in the opposite ocean basin. Some circle the globe.

By the late 1980s NAVOCEANO had the sea level heights measured by the Gedetic Earth Orbiting Satellite (GEOSAT). The application of a high-pass filter to the geoid information gave the ocean floor trends. For those of you who are in doubt, the water level is higher over ocean floor protrusions, such as ridges, seamounts, etc., and lower over depressions, such a geoid lows and trenches. The SASS-based structural diagram for the North Atlantic (Figure 13) was compared to the high-pass filtered GEOSAT data (Figure 14). The comparison was extraordinary; they were exact images of each other for the entire North Atlantic Ocean basin! What this meant was that we had in our grubby little paws the wherewithal to study the structure of all of the ocean basins between 72°S and 72°N. This in itself was a giant step forward because there has historically been very little ship traffic south of the equator on a comparative basis. Large areas, in fact, remain largely unsurveyed. The GEOSAT covered it all, and was not stingy in sharing the info. And, it carried many trends through the sediment ponds, attaching many fractures to seamount chains.

This information has led to a new term, megatrends. Megatrends exist globally. Using satellite altimetry, coupled with bathymetry, these trends go all the way across all of the ocean basins, they lie on at least three different azimuths, and they totally refute any concepts of seafloor spreading. Megatrends are underlain by hot lines, and they are characterized by both earthquake and volcanic activity. They cross trenches, ridges, entire basins, and even continents. Also, they are old, as in, REALLY old. Not any of this 200 Ma for the megatrends, they pass through rocks that are Phanerozoic – Paleozoic in age.

And, the megatrends intersect each other on basin-wide patterns.

Figure 14. GEOSAT high-pass filtered diagram of the Atlantic Ocean basin on a Mercator projection. Comparing this figure to Figure 13 reveals a remarkable resemblance, so much so that one could conclude that the high-pass filtered satellite data is the equivalent of structural trends worldwide. This opens all of the ocean basins for tectonic scrutiny rather than relying on snippets of information gathered in a piecemeal fashion, the provenance of which is open to question in many cases.

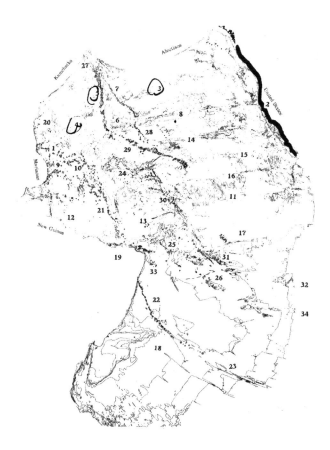

Figure 15. DBDB-5 and multibeam-based bathymetry of the proposed Pacific "plate" at a 1000-m contour interval

on a Mercator projection where: 1=Michelson Ridge, 2=San Andreas Fault, 3=Chinook Megatrend, 4=Shatskiy Rise, 5=Emperor Seamounts, 6=Hess Rise, 7=Emperor Fracture Zone, 8=Mendocino Fracture Zone, 9=Dutton Ridge, 10=Marcus-Wake Seamounts, 11=Clipperton Fracture Zone, 12=Ontong-Java Plateau, 13=Nova-Canton Trough, 14=Murray Fracture Zone, 15=Molokai Fracture Zone, 16=Clarion Fracture Zone, 17=Galapagos Fracture Zone, 18=Udintsev Fracture Zone, 19=Fiji Plateau, 20=Geisha Guyots, 21=Marshall-Gilbert Seamounts, 22=Louisville Ridge, 23=Eltanin Fracture Zone, 24=Mid-Pacific Mountains, 25=Manihiki Plateau, 26=Tubuai Ridge, 27=Krusenstern Fracture Zone, 28=Liliuokaliani Ridge, 29=Hawaiian Ridge, 30=Line Islands, 31=Tuamotu Ridge, 32=Sala y Gomez Ridge, 33=Austral Seamounts, 34=Chile Rise

6-5-1. Pacific Ocean Basin Megatrends

We know that megatrends intersect, so we know that fracture zones do not point in the direction of seafloor spreading: the same piece of real estate can hardly be moving in two different directions at the same time. We can proceed with all due alacrity. The plate tectonic explanation for the central Pacific Ocean basin involves a myriad of twists and turns. A re-contouring effort of the central Pacific basin using all of the available soundings to update that bathymetry produced a *raison d'etre* to reassess Pacific tectonic hypotheses *in toto*. Additionally, the GEOSAT structural diagram also showed previously undelineated trends.

Beginning in the northern basin, the classic definition fracture zones lie on a WSW – ENE lineation. Several are included to make the point. The Chinook Megatrend (Figure 15-3) has been shown to begin in Japan, cross the trench in the form of the Uyeda Ridge, continue through the Nadeshda Basin, underlie the Shatskiy Rise (Figure 15-4), pass through the Emperor Seamounts (Figure 15-5) and Hess Rise (Figure 15-6), cross the Emperor Fracture Zone (Figure 15-7) to become the Chinook Trough, and wind it's way through the sediment traps of the Gulf of Alaska in the form of the three splayed seamount chain/ridges (Figure 10). One could draw the analogy that this has the geomorphic appearance of an eastward-flowing stream, what with the splaying on the eastern extreme.

Just to the south of that is the Mendocino Megatrend (Figure 15-8). This lineament begins on the western Pacific forearc, crosses the trench through the Dutton Ridge (Figure 15-9), continues through the Marcus-Wake Seamount (Figure 15-10) province and the southern Emperor Seamounts as a paleo-fracture, skirts the south end of the Hess Rise, and continues across the rest of the eastern Pacific basin as the Mendocino Fracture Zone, a feature which goes ashore all the way to the Yellowstone National Park. The eastern portion once again seems to splay, encompassing the Surveyor and Pioneer fractures zones.

And, this exercise could be performed for the Murray (Figure 15-14), Molokai (Figure 15-15), Clipperton (Figure 15-11), Clarion (Figure 15-16), and Galapagos (Figure 15-17) fractures too, with the first three seemingly to splay from a common point at the Ontong–Java Plateau.

It would seem as though the eastward splaying is a rule for the megatrends on this lineation, which would necessarily indicate some sort of fluid mechanics at work in the tectonics rather than the movement of a lithosphere "cast in stone" as it were.

I spent many cruises looking at fracture zones in both the Atlantic and Pacific oceans basins. On one cruise in the mid-1980s we kept running into seamounts in the fractures, a phenomenon that was suspected by some, like Rodey Batiza. Most of the cognoscenti believed the fracture zones to be locked, frozen into position as it were. We disproved that. We were checking out the height of one particular seamount, and discovered a nice crater on top. Naturally, being the chief scientist, I ordered a "stop and drop." We dropped several expendable bathythermographs (XBT) into the crater, but found no anomalously high temperatures. I named that seamount after my head compiler, Joe Shanabrough.

As we see in the figure, many lineaments exist on the NNW-SSE axis. Generally, these are thought to be seamount chains created by hotspot activity whereby the Pacific "plate" has moved northerly before 43 Ma, turning abruptly, and almost instantaneously, WNW. This is certainly not the case. In the first place, no major plate realignment happened at 43 Ma. Also, should these trends continue on the same axis, they will eventually have to cross the WSW-ENE megatrends.

Beginning on the SSE, the Udintsev Fracture Zone (Figures 15-18 and 16) actually continues northerly of the proposed Fiji plate (Figure 15-19), passes through the Dutton Ridge (Figure 16) where the Manken/McCann Ridge lies, and ends in the Ogasawara Plateau/ Michelson Ridge at the Izu Trench. This megatrend appears to have been "captured" in the center by the Fiji region, a region which has been hypothesized to be moving eastward by many researchers in a phenomenon called "trench migration." The shoe fits, and I will incorporate that idea.

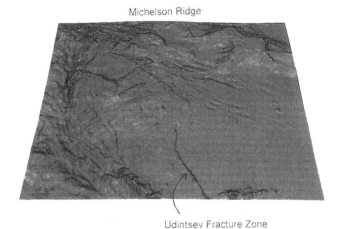

Michelson Ridge

Udintsev Fracture Zone

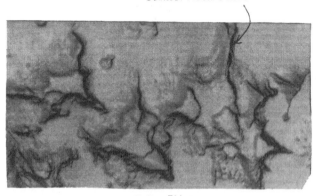

Dutton Ridge

Figure 16. Multibeam sonar-based 3-Ds of the Michelson Ridge (top) and Dutton Ridge (bottom) showing the Udintsev Megatrend.

The Kashima Megatrend lies to the east of that. Beginning on the north in the Geisha Guyots (Figure 15-20), this megatrend splays into two. One passes through the Michelson Ridge (Figure 16), the other just to the east. This double feature continues southerly as the Marshall-Gilbert (Ralik-Ratik; Figure 15-21) seamount chains, is overprinted by the eastern Fiji region, passes into the Lau-Havre Ridge, and reunites in the form of the Louisville Ridge (Figure 15-22) and Eltanin Fracture Zone (Figure 15-23).

The Mamua Megatrend, displaying all the accouterments of a fracture with the perpendicular fabric, crosses the Chinook Megatrend and passes through the Mid-Pacific Mountains (Figure 15-24) as a discrete ridge. From there it is overprinted by the Manihiki Plateau (Figure 15-25), and it loses its integrity after going through the Tubuai Ridge (Figure 15-26).

The Krusenstern Megatrend is the next one to the east (Figures 15-26 and 17). Beginning to the west of the Obruchev Rise at the Kuril Trench, this double-ridged feature becomes a single, downdropped fracture as it passes into the Emperor Seamounts (Figure 15-5). It manifests itself again on the western boundary of the Musician's Seamounts as the Liliuokalani Ridge (Figure 15-28), passes through the Hawaiian Ridge (Figure 15-29), and continues southerly as the Line Islands (Figure 15-30) and the Tuamotu Ridge (Figure 15-31), where it bathymetrically stops somewhere in those sediments.

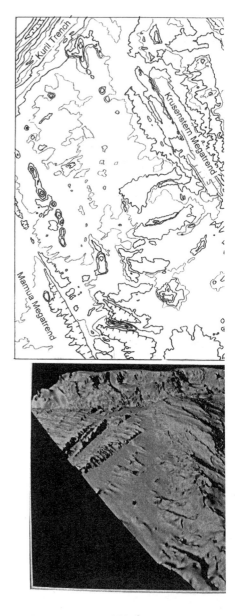

Figure 17. Regional bathymetry at a 200-fm contour interval of north-central Pacific Ocean basin features

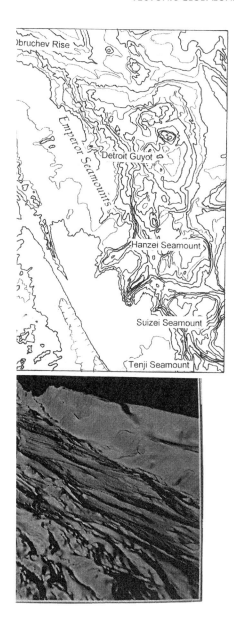

showing the start of the Mamua and Krusenstern megatrends.

I had several experiences with the Emperor Seamounts, having spent a few cruises surveying them. On one trip we were coming back down for a Hawaii inport. The old man turned on the after burners, and we arrived at Niihau about eight hours earlier than planned. You ain't gonna believe this; we went fishing. The DUTTON was about 450 feet long. It cost around $50,000 a day to run it. We tied clotheslines with octopus lures on them to the fantail. The mate put the ship the prescribed distance offshore, and we fished around Niihau for the entire day! Niihau is a volcanic neck in the Hawaiian chain. It is a wildlife sanctuary, or at least it was. You know, we didn't catch one damn fish all day, but the goonies and red-tailed tropic birds were surely nice to watch. I still can't believe it!

The last Pacific basin megatrend I looked at on this azimuth is the Emperor Megatrend. Beginning with the Emperor Fracture Zone (Figure 15-7) on the north, It continues SSE until it merges with the Krusenstern Megatrend at the Liliuokalani Ridge (Figure 15-28 and 18). A possible offset occurs just to the east as the ridge defining the western extreme of the Musicians Seamounts, which fades into obscurity both geophysically and bathymetrically at the juncture with the Hawaiian Chain.

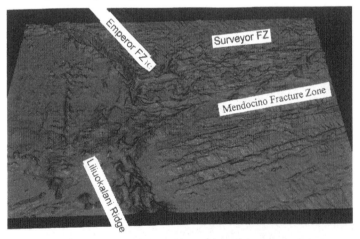

Figure 18. Regional geomorphology of an intersection where the Surveyor and Mendocino fracture zones intersect the Krusenstern/Emperor Megatrend (top). The Emperor portion, after a slight offset, continues SSE as the Liliuokalani

Ridge. To the east (center) of that the Mendocino Fracture Zone splays eastwardly into seven fractures before (bottom) anastomosing to continue into the United States. The bottom portion also shows several ESE megatrends crossing the Murray Fracture Zone, with FFF aligned perpendicular to that in the left-lower corner of that 3-D.

These megatrends do not appear to be splaying at the eastern distal ends, which may be indicative of the age of the features comprising the megatrend.

Several WNW-ESE trending megatrends became apparent when I studied the GEOSAT diagram for the Pacific basin (Figure 19). The first is called the North Pacific Megatrend (NPM), and the second is called the Central Pacific Megatrend (CPM).

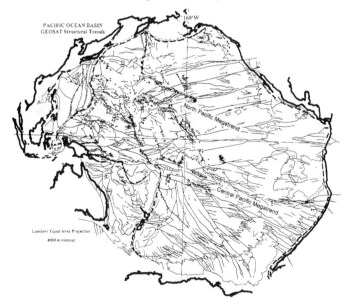

Figure 19. High-pass filtered GEOSAT structural diagram of the Pacific basin based on satellite altimetry data. The larger basin-transecting WNW-ESE trend is the Central Pacific Megatrend. The North Pacific Megatrend can be seen to the north of that.

The North Pacific Megatrend begins from Central Japan where the megatrend is characterized by a conspicuous NW-SE geanticlinal high consisting of Paleozoic rocks. From there the trend slightly changes direction to WNW-ESE, and a continuous ridge is traceable toward the deep sea, south of Daiichi Kashima Seamount. The ridge separates the deep sea trench into two segments; Japan Trench in the north, and Izu-Ogasawara Trench in the south.

Crossing the trench we reach the Geisha Guyots (Figure 15-20), or Japanese Seamounts as they are sometimes referred. Daiichi Kashima Seamount at 35°50'N latitude and 142°40'E longitude, has been proven to be a tectonically undisturbed guyot with Middle Cretaceous (Aptian-Cenomanian) reef at the top. Please note that the seamount chain, although generally getting younger to the east, is not sequentially aged.

The Geisha Guyots have historically not been linked to anything else in the North Pacific Ocean basin, possibly because of the gaps between the features. These gaps also occur in the more famous Hawaiian Ridge and the Emperor Seamounts, so this is not a problem. However, they did not just mysteriously appear on a 120°/300° azimuth some 100 Ma, only to remain anonymous. On closer inspection, they possibly join the Hawaiian chain through the Mapmaker Seamounts.

After a short, buried segment, the megatrend emerges again as the Hawaiian Ridge (Figure 15-29). The Hawaiian Ridge begins with Colahan Seamount, the site through which the Mendocino Fracture Zone/megatrend passes. The Hawaiian Islands are shield volcanoes and shield volcano clusters. They are quite a bit larger than the Geisha Guyots. In fact, the seamounts/guyots all along this chain continue to get larger until the island of Hawaii.

From the Hawaiian Ridge, the NPM crosses the Clipperton and Clarion fracture zones, where the scarce bathymetry does not allow us to follow the passage. The GEOSAT data (Figure 19) shows a rather prominent acute strike crossing those fracture zones to enter into the realm of the East Pacific Rise. In light of the ground-truthing of all the rest of the GEOSAT features, we will assume that this is our NPM and continue with the description.

TECTONIC GLOBALONEY

Before continuing to the east between the East Pacific Rise and Central/South America, we need to catch up with another megatrend which will blend into this on, the Central Pacific Megatrend (CPM). The massive CPM begins for the purposes of this discussion the Banda Sea surrounded by five crustal elements: the Sunda Shield (Carboniferous age), Indian oceanic crust (Cretaceous), the Australian craton (Paleozoic), Pacific oceanic crust (undetermined), and a transitional complex (Paleozoic).

The CPM begins as a single 3-4 km-tall, 100 kilometer-wide ridge continuing out of the Banda Sea through New Guinea (the Maoke Mountains) and the Ontong-Java Plateau (Figure 15-12), across the northern Fiji Plateau (Figure 15-19), and through the Samoan Islands for a distance of over 4000 km. This megatrend continues as the Tuamotu Ridge (Figure 15-31), the Easter Island Fracture Zone, and into the East Pacific Rise, where it splays into three segments.

The northernmost fork joins the flow at the East Pacific Rise to go northerly and join the NPM to become the Galapagos Ridge (Figure 20), which itself bifurcates into the Cocos Ridge on the north and the Carnegie Ridge on the south. The Cocos Ridge obliterates whatever trench may have been there and abuts Costa Rica. Then the ridge passes through Costa Rica to emerge into the Caribbean Sea as the Hess Fracture Zone/Ridge supporting the various islands starting with San Andreas and continuing through Jamaica and Haiti, a ridge between the Caribbean Sea proper on then south and the Cayman Trough on the north. The southern segment includes the Galapagos Ridge and Carnegie Ridge. These features are supposed to be the divergent boundary between the Cocos and Nazca plates. The islands, also hypothesized to be a hotspot track, show a mixture of ages.

Several more ridges appear in the bathymetry. The Coiba Ridge trends more northerly, rises about 3000 m, and is a relatively short 200-km in length. The Malpelo Ridge, also arising from nowhere, passes northeasterly for about 400-km before fading into oblivion. It is broader but of lesser extent, rising only 1500 m.

Somewhere in this region in the older bathymetry was supposed to be a plate boundary, with the foregone descriptors applying to the Cocos "plate." We can see no reason to suspect such an event based on the updated bathymetry.

Additionally, from the Cocos Ridge to the south of the Carnegie Ridge, no trench exists in the bathymetry. The regional base depth appears to be about 3500 m, and several 4000 m readings occupy 2°30'N, 84°30'W and 6-7°N, 76-77°W. Subduction has been theorized for this region for the past 75 Ma. The earthquake regime is primarily above 200 km in depth, with no earthquake activity below 250 km. This means that no subduction can possibly be taking place from the Cocos Ridge on the north through 5°S, or from the Tehuantepec Ridge at 15°N to 5°S, at least for time covered by the data collection process.

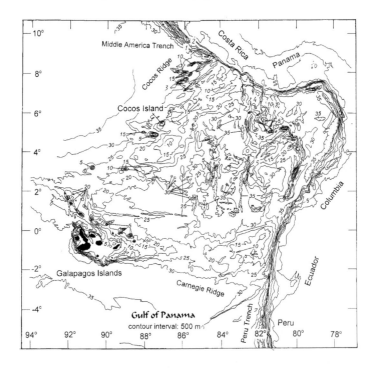

Figure 20. Gulf of Panama/Cocos "plate" at a 500-m contour interval and Mercator projection, compiled from ship-of-

opportunity NGDC data base with an insert around the Galapagos islands from Bill Chadwick's 1994 chart. This is the region where the megatrend is joined by those from the north and south East Pacific Rise to accentuate the effects of the magma coming in from the WNW. The Cocos and Carnegie Ridges are the primary bifurcation. With compression acting on this feature, the (4) Coiba and (5) Malpelo Ridges form. Although they do not appear at this contour interval, (1) the Fisher Ridge and the (2) Fisher Seamount have anomalous ages. (3) Quepos Plateau has also been dated.

Dong Choi has extensively researched western South America and its relation to the subduction process (*Active Margin Tectonics*, 2001). The Carnegie Ridge goes into the South American continent at least through the Guyana Shield. In a study of the seismic data, major Precambrian (Proterozoic) structural trends in the South American continent clearly continue into the SE Pacific. Published geologic and tectonic maps show that major Precambrian (particularly Proterozoic) structural trends recognized in the Precambrian shield connect to the main bathymetric features of the SE Pacific Ocean basin (Figure 21; with permission). Although the connection is somewhat disrupted and obscured in the Phanerozoic circum-Pacific mobile belt area, the connection between the shield and the ocean is obvious. Even within the mobile belt area, overall structural highs follow the trends of the shield. Precambrian rocks are exposed north of Santiago, where a large NW – SE-trending Proterozoic anticlinal axis extends from the shield. The Precambrian rocks extensively occupy the areas both south and north of Lima where the Guyana and Brazilian Shields extend. No Precambrian rocks are exposed in southern South America.

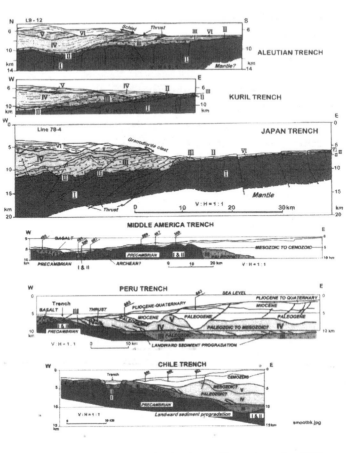

Figure 21. Reinterpreted seismic profiles across the NW Pacific trenches and the Central and South American trenches. Note, 1) well developed block faults in Units I and II in all profiles, as well as anticlinal structures in the trench areas in the bottom three profiles, Central and South America, 2) sediment progradation pattern observed in Units III and IV toward the present continents implying the presence of provenance in the areas presently occupied by deep ocean during the time of deposition of these Units (Paleozoic to Mesozoic?), and 3) thrust faults in the lower continental slope in many profiles. Numerous continental

rocks have been dredged in the Japan and Kuril Trench areas. As clearly seen in all profiles, no subduction is occurring. (Used by permission of Dong Choi)

The above facts suggest that the major submarine features appear to be directly controlled by the Proterozoic structures, and that the real oceanic basement under the younger oceanic basalt may consist mainly of Proterozoic rocks instead of the Cenozoic basalt. This alternative view is supported by paleo-geographic and sedimentological studies, which established the presence of paleo-lands in the present Pacific Ocean basin during the Paleozoic, the Mesozoic, and into the Paleogene on sound bases. Many of the sources consider that the paleo-lands have disappeared due to subduction. Last, the eastern equatorial Pacific has been defined by the plate tectonicists as one of several plates: the Cocos and the Nazca. The Cocos Ridge has been proposed to be the buoyant trace of the Galapagos "hotspot." and that the ridge is presently subducting the trench at the rate between 7-9 cm/yr and 22.5 cm/yr. One has to wonder where this information could possibly arise as no trench exists in this region. The eastern margin of the ridge is truncated by the Panama Fracture Zone, a feature dividing the Cocos and Nazca plates. Additionally, the Cocos Ridge appears to arise to the west of the Galapagos "hotspot" site.

Therefore, no reason exists to have this rather complicated tectonic explanation for events in this region. In fact, nothing exists having even the remotest appearance of a spreading center, or of a propagating rift, or of a subduction zone. Rather, the region appears more in the bathymetry as an eastward-pointing delta, a feature showing stream flow characteristics. Here again, this geomorphology is descriptive of a hot line, especially with the typical delta flaring on the eastern edge.

Finishing the CPM, the central fork continues as the the Sala y Gomez Ridge/Easter Fracture Zone (Figure 15-32). A study has carried that lineament onto the South American continent through the Nazca Ridge and onto the Precambrian Brazilian Shield, even though the sketchy bathymetry shows a subducting feature (Figure 22). The third, and southernmost, fork of the CPM starts on the south of the Ontong-Java Plateau, follows the same path, but splits off at Samoa to the south. It is comprised of the Austral Seamounts (Figure 15-33) coming out of the Fiji Plateau with Machias Seamount, the southern portion of the Society Islands, and it continues through south fork of the Easter Fracture Zone, the Juan Fernandez structure on the East Pacific Rise, the Chile Rise (Figure 15-34), South America, and the Falkland Islands in the south Atlantic Ocean. This megatrend passes through Precambrian rocks.

Figure 22. 3-D of the NGDC ship-of-opportunity data as of 1998 of the intersection of the Nazca Ridge with the Peru (top) and Chile (bottom) trenches. Vert. Exag. X5

These megatrends all seem to splay on the eastern extremes, and they all appear to be the only active megatrends in the Pacific basin. So, they are at once both extremely young and extremely virile.

In the Introduction you were shown an extraterrestrial megatrend on Venus (Figure 1), a sister planet. In *Uncovering the Secrets of the Red Planet: Mars* (National Geographic Society; 1998), a sister planet in the other direction, the introduction there reads like it belongs here. The size of Mars has allowed the tectonic regime free room to roam throughout much of its 4.6 Ga history, tectonic processes presumably driven by a heat engine. The highland-lowland boundary is called the Tharsis bulge, and it rises 10-km above the datum (seasurface for Mars) containing four enormous volcanoes with many smaller ones (starting to sound familiar?). It is contained in a system of tectonic features such as rifts and grabens, which are long valleys formed during extensional processes. Valles Maneris is the longest at 4000-km. Wrinkle ridges, thrust faults and folds characterize this feature. Just as on Earth, these features seem to have "resulted from adjustments in the outer layers of the planet due to the formation of the Tharsis bulge. Are we talking about the CPM or Mars here? They sound exactly the same. Compare Figure 1 with the Pacific GEOSAT diagram (Figure 19). Presumably, as the Mars information is digested more robustly, the bulge will appear as the two preceding features.

6-5-2. Atlantic Ocean Megatrends

The alignment of Atlantic basin fractures has always been generally thought to be EW, so I assumed this to be the case when analyzing the superchart.

Not so; Art Meyerhoff and I did a study of Atlantic fracture zones in 1995. Thirteen transform faults were found to cross the Mid-Atlantic Ridge between 0 and 55°N latitude. The fracture zones are absent from 40°N to 52°N between the Hayes and Charlie-Gibbs fracture zones. An additional 25 to 38 fracture zones lie off-ridge (Figures 13 and 14). Hence, fracture zones do not necessarily form as extensions of transform faults, apparently being controlled more by the regional stress field. Also, the idea that they may have been formed before the implantation of the midocean ridges began to take shape.

Somewhere along the line I was on the *BOWDITCH* doing a survey in the Bermuda Triangle/Sargasso Sea region. We were trying to plant a triad of transponders to survey this one particular area. In the old days, we didn't have GPS; we actually had to know how to do something that a computer couldn't do for you. Interesting concept, huh? Anyhow, these transponder had a particular signal for interrogation by the on-board chirper, and also a particular sequence at that frequency which would release the buoy from its mooring, for example, a 12.5 kHz ping would be answered and received back on board. The two-way travel time would then be figured from each of the three, and a triangulated position would be obtained. A bunch of satellite fixes were also being *gathered during this survey, and they were used to establish the real position of the survey. The transponder work was very accurate because we took an ocean station using Nansen casts to determine the speed of sound through seawater at that spot for that time. OK, 'nuff said. We were enjoying this huge school of porpoises while being out on deck. As we tried to position the transponder triad, the depth figures kept getting shallower. This was not possible, until the beacons started popping back up on the surface. We couldn't believe it; the porpoises had scoped in on the exact frequencies of each transponder and, just like the room full of monkeys, they had released all of the transponders. After about four tries, we had to sail away for about a week and then come back to work that area.* Evidence for sea or land bridge

In the synopsis of the available data of the Atlantic Ocean basin, most of the megatrends are aligned WNW – ESE in the North Atlantic. Starting on the north at the Charlie-Gibbs Fracture Zone and continuing south to the Azores/Pico Fracture Zone, almost no E-W fractures exist. Instead, this region is characterized by a high number of seamounts; nearly 200, according to the superchart, in all with no particular alignment. Nowhere else in the Atlantic Ocean basin does this phenomenon occur. Nor, do they reach the midocean ridge crest. The possibility remains that, although seamounts do show an underlying fracture in most instances, this region has a different stress field because the ocean basin is so narrow between the continental cratons.

From that point to the equator, three of the major fracture swarm/megatrends traverse the entire basin (Figures 13 and 14). The Oceanographer/Hayes/Atlantis swarm includes the seamounts at the distal ends near the continents. The second is the megatrend comprised of the Barracuda/Vema/Guinea fracture swarm and includes the Bathymetrists Seamounts. This feature offsets the midocean ridge for the distance of 1100 km and appears to go ashore in western Africa on the east. The western terminus on the Atlantic Ocean floor appears to be the Lesser Antilles island arc. The third is the St. Paul/Romanche fracture swarm, which straddles the equator to 3°S latitude (Figure 23-7). It is here that the megatrends change direction from WNW – ESE to WSW – ENE, and this trend becomes more pronounced the further south one looks. Fracture trends in the South Atlantic basin are generally evenly spaced and do not display the contortions provided by the counterparts in the north.

For the Bullard fit to work, Massachusetts, USA, must fit with Morocco, Africa, and fracture zone(s) must join the two. This is the plate tectonic definition. The Hayes/Oceanographer fracture zone swarm, which included the attendant seamounts, spanned the region from Ontario, Canada (which lies mid-continent), through Massachusetts USA, across the North Atlantic basin, and on to the Iberian peninsula was a surprise. Those fractures had never been noticed to join. On the western end of the fracture swarm the New England seamounts had been noticed to go ashore through Vermont to the Monteregian Hills in Quebec. The New England Seamounts/Monteregian Hills were also known not to be age sequential, ranging in an admixture from 70 to 230 Ma. They lie on substrate that is at least 1 Ga. On further investigation, the trend continued ashore as the Betic Cordillera crossing southern Spain. The basic meta-sedimentary rock there is Permian-Triassic in age. The fault zone associated with southern Spain is thought to be of a vertical nature rather than strike-slip. The megatrend is also in line with the Ballaeric Islands in the Mediterranean Sea.

Nevertheless, the WNW-ESE trending north Atlantic fractures exist, and we see the WSW-ENE trending south Atlantic fractures (Figures 14 and 23). Interestingly, this fact remains largely uncontested. However, this alone belies the concept of unidirectional Atlantic basin spreading.

On the structural diagram based on bathymetry, trends paralleling the midocean ridges are also noticeable. One starts on the Bermuda Rise, passes NE through the southern Sohm Plain, skirts the eastern portion of the Corner Seamounts while crossing the Oceanographer/Hayes megatrend, is overprinted by the Mid-Atlantic Ridge, goes into the King's Trough region, and continues into Ireland, across the Irish Sea, and on through Scotland. I have not studied any further north, merely noting the existence of this trend, called the "Caledonian" in Scotland. Presumably this feature formed on land during the Caledonide Orogeny, a period which lasted from the Ordovician (about 505-438 Ma) to the Devonian (began about 409 Ma). The Bermuda Rise is known to be continental in origin with DSDP ages of 671 (Proterozoic) and 381 (Devonian) Ma. As noted, this trend also passes through the

King's Trough. Ordovician (440-530 Ma) graptolites and trilobites have been found there. The King's Trough is on the ocean floor, so we may take the liberty of assigning this age to the entire feature, and to herein name it the "Caledonian Megatrend."

Figure 23. The African "plate" at a DBDB-5-based 1000-m contour interval on a Mercator projection where: 1=Cruiser/ Hyeres seamount group, 2=Madeira Island, 3=Canary Islands, 4=Atlantis Fracture Zone, 5=Cape Verde Islands, 6=Kane Fracture Zone, 7=Romanche Fracture Zone,

8=Ascension Island, 9=Ascension Fracture Zone, 10=Bode Verde Fracture Zone, 11=Bagration Fracture Zone, 12=Rio de Janiero Fracture Zone, 13=Tristan da Cunha, 14=Cape Basin, 15=Madagascar Ridge, 16=Reunion, 17=Madagascar, 18=Mascarene Plateau, 19= Carlsberg Ridge, 20=Red Sea. Supplemental contours were provided by a blending of the 1989 GSA-published "Bathymetry of the South Atlantic Ocean." The reader will notice that this "plate" is almost totally surrounded by "spreading centers."

Similar SW-NE trends appear in the south Atlantic, one encompassing the northern cluster of seamounts on the Guinea Rise (Figure 23), and one underlying the Walvis Ridge to the south (Figure 23). Oil exploration reports on-line show four stratigraphic units for the Walvis Ridge and it's onshore counter-part, Namibia. The pre-rift Karoo section contains preserved Permo-Carboniferous-Triassic age samples, and this section continues offshore. The Walvis Ridge is a later addition to this base.

Last, several NW – SE trends appear in the GEOSAT structural map. The swarm on the western terminus of the Kane Fracture Zone continues southeasterly and is co-linear with the eastern trends between the Barracuda and Vema fractures. Another appears at the coast of Brazil, about 25°S latitude, 40°W longitude and continues trans-basinal to 50°S latitude, 20°E longitude.

6-5-3. Indian Ocean

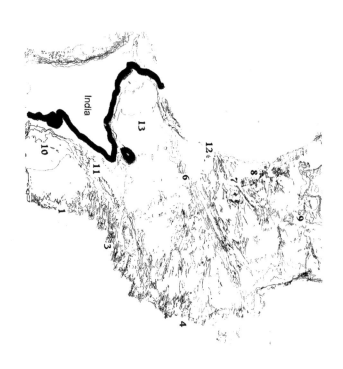

Figure 24. Indian-Australian "plate" based on DBDB-5 bathymetry updated by use of GEOSAT at a 1000-m contour interval on a Mercator projection where: 1=Carlsberg Ridge, 2=Owen Fracture Zone, 3=Mid-Indian Ridge, 4=Southeast Indian Ridge, 5=Diamantina Fracture Zone, 6=90-east Ridge, 7=Investigator Fracture Zone, 8=Indoysian Foldbelt, 9=Broken

Basin Megatrends

Ridge, 10=Arabian Sea, 11=Chagos-Laccadive Ridge, 12=Andaman-Nicobar Trench, 13=Bay of Bengal, 14=Australia-Antarctic Discordance, 15=Banda Sea, 16=Lord Howe Rise, 17=New Hebrides Trench, 18=Fiji, 19=Tonga Trench, 20=Lau Ridge, 21=Kermadec Trench, 22=Norfolk Ridge, 23=New Zealand, 24=Hjort Trench, 25=South Tasman Rise.

The bathymetry (Figure 24) and the GEOSAT data were used to construct a basic outline of events in the Indian Ocean basin. The most glaring observation to be made for the Indian Ocean basin is the differences in strike and pattern of the fracture zones, especially between the SW and NW/SE quadrants. The NW fractures show a left-lateral drag effect in the bathymetry, and this includes the Carlsberg Ridge (Figures 23-19 and 24-1) in the NW quadrant. From the triple junction near 25°S latitude, 70°E longitude to the Owen Fracture Zone (Figure 24-2) at 10°N latitude, 55°E longitude, the fractures trend SW-NE. For the Southwest Indian Ridge (Figure 23), they trend N-S. The fracture spacing is also different. It is closer on the Southwest Indian, Mid-Indian (Figure 24-3), and Carlsberg Ridges and more widely spaced on the Southeast Indian Ridge (Figure 24-4). North and south of the Southeast Indian Ridge fractures do not exist at this contour interval. Given a little assistance by the GEOSAT diagram, the fractures appear to be similar in length-to-longer than their basinal counterparts, and they appear to stop at the Diamantina Fracture Zone (Figure 24-5), a WNW-ESE megatrend.

Observation allows this interpretation. Fractures on the SW branch of the Mid-Indian Ridge appear to have been dragged into the morphology now existing because they point more easterly than do the fractures to the north on the same ridge.

Outside the midocean ridge segment fractures, the large fractures seem to be in the eastern basin. Several large, N-S trending ridge/trough structures lie in that portion of the basin; the 90-east Ridge (Figure 24-6), a seamount chain, and the Investigator Fracture Zone (Figure 24-7). The SW-NE trending fractures do not appear to transect the 90-east Ridge, instead they seem to have been previously imprinted and are now considered to be inactive. That same fracture deflects to the north at the Investigator Fracture Zone. This megatrend, by the way, is now being touted as the next plate boundary by the adherents of the plate-tectonic hypothesis. That is based on the enormity of the mid-basin earthquakes.

The Indoysian Foldbelt (Figure 24-8) and the Broken Ridge (Figure 24-9) are the sites of the E-W trending fractures. The fractures between the Diamantina Fracture Zone and the Broken Ridge do not cross either of them. On a smaller scale, E-W – trending fractures appear around the Kerguelan Seamount Chain, and they appear to be co-linear with Diamantina.

The fractures to the south of Africa are aligned SW-NE and would appear to be extensions of those to the east of Madagascar were that island not in the way. The fracture pattern in the Arabian Sea (Figure 24-10) gives the impression of a whorl, where they intersect the Chagos-Laccadive Ridge (Figure 24-11) on the east, bend to the SW to intersect the Carlsberg Ridge, and trend generally NE in the Gulf of Oman. Another whorl appears to the south of the Andaman-Nicobar Trench (Figure 24-12) off NW Australia. The whorls do not coincide with gravity geoid highs or lows.

Seamount provinces appear, once again at this contour interval, to be mostly off-ridge. The primary sites are at the Owen Fracture Zone, the Chagos-Laccadive Ridge, north and northeast of Madagascar, south of the Bay of Bengal (Figure 24-13) and north of the Diamantina Fracture Zone. In the Indoysian Foldbelt, the seamounts are aligned E-W, and they seem to merge with those on the 90-east Ridge. On-ridge seamounts are very scarce, even on the GEOSAT structural map. The larger ones appear on a N-S azimuth on the 90-east Ridge. Two smaller chains appear to the east and are associated with the Investigator Fracture Zone. An even smaller number appear to the east of that, and they may be an extension of those at the trench segment which included the large volcanoes such as Krakatau. The Indian Ocean does not have many seamounts, possibly because the ocean basin is filled with oceanic ridges. Most of the available magma is tied up the trunk channels.

Truly, the megatrends in the Indian Ocean basin seemingly invite descriptions such as "chaotic, jumbled, and confused". Five sets of megatrends exist in the Indian Ocean basin. Obviously some, some of these megatrends must cross each other, especially the trans basinal ones.

6-5-5. Marginal Seas

Figure 25. SE Asia marginal seas where the South China Sea is presented at a 1000-m contour interval on a Mercator projection and the rest is from the GEBCO chart at 1000, 2000, and 4000-m contour interval where: 1=Flores Sea, 2=Banda Sea, 3=Weber Deep, 4=Tanimbar, 5=Irian Jaya, 6=Seram, 7=Halmahera, 8=Celebes Sea, 9=Sulu Sea, 10=Philippine Trench, 11=North Borneo Trough, 12=Spratly Islands, 13=Palawan, 14=South China Sea, 15=Paracel Islands, 16=Manila Trench, 17=Luzon Strait.

While we're showing bathymetry of the world's ocean basins, we may just as well include the marginal seas, such as the South China and Phil Sea regions (Figures 25 and 26). No discernable fracture zones were found there, but trends abound, albeit rather difficult to create a megatrend in an enclosed sea. However, the overall SW-NE alignment of the South China Sea is very important in a later discussion, so we will remember the Spratlys (Figure 25-12), the Paracels (Figure 25-15), and the unnamed chain of seamounts in the center of the South China Basin (Figure 25-14)

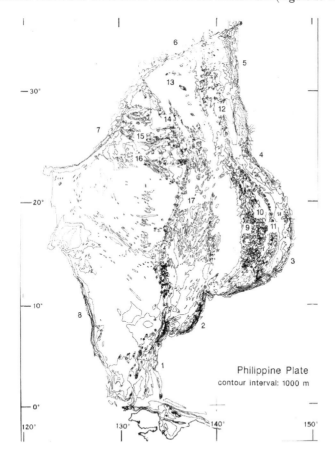

Figure 26. The Philippine "plate" at a 1000-m contour interval on a Mercator projection where: 1=Palau Trench, 2=Yap Trench, 3=Mariana Trench, 4=Bonin Trench, 5=Izu Trench,

6=Nankai Trough, 7=Ryukyu Trench, 8=Philippine Trench, 9=West Mariana Ridge, 10= Mariana Trough, 11=East Mariana Ridge, 12= South Honshu Ridge, 13=Shikoku Basin, 14=Kyushu-Palau Ridge, 15=Daito Ridge, 16=Oki-Daito Ridge, 17=Parece Vela Basin.

From this section we may add to our working hypothesis intersecting fracture zones/megatrends and vortex structures. Thus, the micro-cracks remove the micro-plates from the working hypothesis.

6-6. Geophysical Superswells *vs* Bathymetric Contours of a Region

Plate Tectonic Presumption: Bathymetric superswells inhabit the ocean floor in a random fashion, one of which encompasses the entire western Pacific basin. Harry Hess, using the guyot heights he had collected during World War II, hypothesized a large western Pacific "superswell." This type of feature is created by the rising of lots of magma in a mantle plume, like a super hot spot. Mantle plumes really caught on to help explain the "Great Cretaceous Outpouring," an event during which most of the western Pacific seamounts formed.

Actuality: We are immediately served with several items requiring further explanation before we continue to the actual bathymetry, that of mantle plumes and hotspots and the formation of Large Igneous Provinces (LIPs), such as the South Pacific Superswell. Hotspot activity is predicted to build seamounts and islands. The Society Islands, Tubuai Ridge, Tuamotu Ridge, Line Islands, and Marshall-Gilbert Islands/Seamounts are integral "proofs" of the original idea proposed by Morgan. The rises and plateaus are mantle plume creations. Mantle plumes are theorized to have built the continental flood basalt regions, which average 330,102 km^3/feature based on 25 that have been measured. The oceanic plateaus are 15.22 x 10^6 km^3 feature based on 34 samples. The Deccan Traps formed at the intersection of three rifts for future reference.

Anyhow, Menard took the ball and ran with it. Menard had the entire western Pacific basin to be a superswell (1964), called the Darwin Rise. By 1984 he had realized the error of his ways, presumably because he finally looked at the bathymetry. Menard removed the superswell south of the equator from his mythical feature. However, Menard died, and some of his students floundered in the ocean depths without guidance. These, followed by their students, have dug themselves in even deeper. Let us take a look at the actual bathymetry. After all, if a swell exists, it will certainly appear as elevated contours.

To exonerate Menard's decision to omit the southern portion of the Darwin Rise, the route of the CPM, shown above, clearly defines the regional base depth at 5400 m. This is the depth on the west, in the center, and on the east, all the way across. No sign of raised contours exists with the exception of the CPM ridges. This exercise has reaffirmed Menard's original deletion of the southern portion of this superswell from the ranks of anomalous megafeatures.

After I had written a few papers on guyots, I tried to summarize the findings in a conclusive paper. The journal editor sent it to Bill Menard to review. In what was probably the most egotistical, self-serving review I ever had the misfortune to receive, Menard took credit for everything all the other investigators, especially Harry Hess and I, had discovered. He tried to claim that he had done all of the work. I replied to the editor that Menard did not have access to the multibeam sonar data that I did, so where did he come off with such hubris. Menard had a bigger name, so he won out. I had to get the paper published in a less well-known journal. Needless to say, I reamed him a new one the first, at last, time I ever met him. Because he called a paper "The Darwin Re(pri)se," I later named a paper "The Darwin Rise Demise."

That still leaves the dream-feature in the North Pacific. An exercise plotting guyot heights showed a different scenario than a swell-like feature; it showed a fracture zone instead. As we saw in the last section, the CPM refutes totally the idea of a South Pacific Superswell for once and for all. Hopefully no more tax dollars will be spent in that direction, but one never knows about such things. Stranger things have happened. Coupled with the disproving of the Darwin Rise, one wonders where the idea of a superswell arose in the first place. One certainly does not exist in the Pacific Ocean Basin.

The plate-tectonic hypothesis explains intra-plate anomalies by hot spots, or fixed mantle plumes, where linear seamount/island chains show an age progression away from the hot spot. Very few age progressions exist in the real world to verify this concept.

On further investigation, not one of the seamount/island chains listed above has an age sequence, so we are not going to worry about hot spots. Lately, different investigators have independently found no evidence of anything resembling mantle plumes (such as Iceland and the Deccan Traps), hotspot tracks, or age-sequential seamount chains. In fact, "linear seamount chains delineate the stress field, not the displacement field." They are formed more by a "hot line."

If the very idea of thermal mantle plumes has now been totally discredited, one may ask with a minimum of trepidation: Why are we being led down the primrose path of hotspot/mantle plume origin for the plateaus and rises? Clearly some other means of construction has been at work here that has not been considered. What is relevant is the overall alignment, the location of the earthquakes, and the total agreement of the structural trends provided by the GEOSAT to the bathymetry. All the larger ocean floor features reside atop at least two megatrend intersections; ALL OF THEM.

6-7. Paleobiogeography and Gondwanaland

Fossils, or, more specifically, Early man fossils, have been an ongoing hobby of mine since I was a kid. Some kids like dinosaurs, some like early man. Anyhow, I have been paying particular attention to the trek "out of Africa." We have been bombarded with justifications about how we evolved in Africa about 2 Ma and moved out to populate the world in the popular literature. After all, most of us do not have access to, nor are we even aware of, the publications of the American Geophysical Union for example. We see the Discovery and A&E channels on TV and read National Geographic, Discover, and Scientific American. The early mover and shaker is called Homo erectus, or upright man. His presence is once again based on fossils (rock samples). You all saw how rock samples had and have been misrepresented to the paying public and students alike. Well, friends and neighbors, it's happening again.

A study of mitochondrial DNA by Sarich and Wilson in 1987 "proved" that we all came from a 200,000 year old African, a "mightychondrial Eve" as it were. The fact that one of Wilson's co-workers, Rebecca Cann, later stated the fallacy of this study has meant nothing to the prehistoric out-of-Africa group, a group which seem to hold most of the funding.

The oldest erectus fossil found in Africa is 1.0 Ma. However, erectines have been found all over the Old World. They have a couple in Indonesia dated at 1.8-1.6 Ma. They have several in China dated at 1.78-1.96 Ma. One has been found in India dated at 1.4 Ma. Many, along with tools, have been found in Soviet Georgia at 1.7-1.2 Ma. Starting to get the message yet? How can a Homo erectus who is only 1.0 million years old move backwards in time 0.8 million years to populate the Far East?

The addition of the fossils, very well preserved, from Dminisi in Georgia (SW Asia), dated at 1.6 Ma, has brought exclamations such as "put them back in the ground." The skull cases are primitive, the braincases are smaller than those of the erectines. The out-of-Africans have given us the 2.0-1.0 Ma Homo ergaster in Africa. Now they want to do away with that classification and lump all of the post-Homo habilis specimens into the erectus category. That way we can still come from Africa and not have to upset the funding agencies too much.

Additionally, both the Chinese and the Australian anthropologists have enough remains to show a straight-line descent for the modern Australian aborigines from those early Indonesians to show very little variation whatsoever. The Willandra Lakes 50 skull is about 50,000 years old. The Cossack skull is only 6500 years old. Skulls in the direct line are also found in Cohuna (undated), Talgai (9-11,000), Kow Swamp (13,000), and Mossgiel (see Archaeology of Dreamtime, Josephine Flood, 2000). The Aborigines do not think that their ancestors came from Africa, and they know a whole lot more about it than we do. Almost looks like man evolved in Asia and moved backwards in time to get to Africa, leaving poor old Australians out in nowheresville.

However, we are currently bombarded with journal/magazine article titles such as: "The African Emergence of Early Asian Dispersals of the Genus Homo" (American Scientist, Vol. 84, 1996), "Out of Africa" (Earth, 1996), "Ancient Roads to Europe: African Ancestors May Have

Entered Europe Surprisingly Early" (Science News, Vol. 151, 1997), "Out of Africa Again – and Again?" (Scientific American, 1997), "Fossils Hint Who Left Africa First" (Science News, Vol. 157, 2000), and finally "Not Out of Africa" (Discover, 2002). You get the picture.

Actually, trying to make things more agreeable, both sides now seem to be moving towards the center by calling all prehistoric men Homo sapiens. Interesting; we'll have to sty tuned to see where this goes. Look out Neanderthals.

Plate Tectonic Presumption: Based upon, among other things, continental drift was demonstrated to us in part by the distribution of the fossilized *Lystrosaurus, Glossopteris,* and several other indicators. India is known to have moved northerly and smashed into Asia, thus causing the Himalayan Mountains to rise.

Actuality: We begin with rocks, or, in this case, fossils. Paleobiogeography, a big word meaning "where the plants and critters were many years ago," was used to demonstrate the inner workings of tectonics. The premier paleobiogeography study was published in 1996 by the Geological Society of America, a flagship of geoscience on the American continent. The study was performed by Arthur Meyerhoff, Art Boucot, Mac Dickins, and Donna Hull, which, of course, immediately raised a red flag among the scientific cognoscenti. What Meyerhoff and his crew did was plot fossil localities, both temporal and spatial. What they found was an intercalary zone where the northern and southern taxa were intermixed in the same bed from a line at approximately 50°N to 50°S. This shows that the continents have not moved in relation to each other since the Proterozoic with the possible exception of E-W movement.

By this time I had met Art Meyerhoff. He had such respiratory problems that he was unable to attend his valued conferences any more or visit with his scientific-minded friends. He didn't like telephones, so he had a secretary type all of his correspondence. I still have a book full of letters we traded back and forth where he talked me into getting back into the publishing world. I knew that the plate tectonic paradigm did not work, as you have seen herein, and was having a horrible time of trying to get anything through the reviewers. I did not get paid to "publish or perish," so I just did my 8-5 and was *minding my own business. I was sitting in Art's living room in Tulsa when the reply to his monograph came from the publisher's office at GSA. He said that he did not want to publish this treatise because of who Art was, but that he could not ignore the science because it was all true. When GSA Memoir 189 came out, part of the preface was: "Biogeographical data provide a potentially powerful tool for deciphering the tectonic evolution of the Phanerozoic Earth because the borders of biogeographical units are natural barriers, some of them tectonic in origin ... do not always coincide with plate boundaries proposed by many geologists and geophysicists. This discrepancy exists because most scientists who work on global tectonics rely mostly on geophysics, tectonophysics, and structural geology ... The questions raised by the presence of this intercalary zone are complex, and not easily solved ... In the case of plate tectonic models, physicists and geologists, by ignoring the biological disciplines, only weaken their own arguments ... "*

Let's assume that we didn't already know about the studies of the Wadia Institute of Himalayan Geology; let's assume that India had a nice trip across the Indian basin for about 50 Ma. We have a mobile India, docking with the Eurasian continent a mere 15 Ma and raising the Himalayan Mountains. This mobility/orogenesis is based on tetrapods, invertebrates, floras, stratigraphy, and magnetics. Proximity has also been proven by Late Cretaceous-Danian terrestrial vertebrate faunas. India, and tiny *Lystrosaurus*, are integral parts of Gondwanaland. *Lystrosaurus* has been found north of the Taurus-Zagros-Indus-Yarlung suture zone; that is, in Asia proper. A study mapping all of the *Lystrosaurus* remains included Antarctica, Africa, the Moscow basin in Russia, India, western China, and Vietnam. The sample base increased, but to date none have been found in the Americas. In fact, four amphibian families have been found in the same bed only in Antarctica, South Africa, and Tasmania. All four are found from as far away as Asia and Svalbard, but nowhere else are they in the same bed. Speculation now has it that *Lystrosaurus* evolved in the north and migrated south during the Triassic. Ergo, Lystrosaurus cannot be used to prove anything about drifting, and India has been found to be close to Asia and even Africa, but never close to the Gondwanaland continents.

That leaves India somewhere in the Indian Ocean between 65-15

Ma. Among other things, the magnetic anomalies associated with the mid-Indian Ocean ridges have been misinterpreted. As we saw in section C, studies based on rock samples collected *in situ* tell a different story. The fact that India has not gone anywhere will be important at a later discussion, so it has been here noted for reference.

Australia is another case in point. During the Triassic, Australia was largely isolated. The *Kannemeyeria* (*Cynognathus*) Zone fossils are only in Australia during the Early – Middle Triassic on through the Cenozoic. Antarctica was isolated as well. Another real proof lies in the fact that many dredge hauls on the Australia-Antarctic Discontinuity (supposed spreading ridge between the two continents) yielded nothing but granites. Both Australia and India are part of the fabled Gondwanaland. Antarctica has not been in close proximity to anywhere except for southern South America.

Crocodile bones have been found in both Antarctica and Greenland. As crocs usually frequent the warmer climes, this poses a problem to the non-drifters. However, a mere tilting of Earth's axis of something like 23° will get the polar regions warm enough for *Glossopteris* plants and crocs to live. This much tilt has already been proven by NASA, as the pull of the Moon provides this agent. Based on the lineament trends of the megatrends, this angle of tilt has varied as much as 120° in the past. So, the continents did not have to display any degree of upward mobility.

As far as the ocean is concerned, being that it is still 70% of Earth's surface, nobody has found any dinosaur bones on the ocean floor. Paleobiogeography remains a moot point as to the prior existence of land bridges. However, enough continental-style and old age rocks have been found in the Atlantic basin from the equator north through Greenland that many scientists are claiming that the Atlantic Ocean has always been the size it is now, and that free passage back and forth between Europe and America was a fact. This is true for the land bridge through Greenland/Iceland as well as the one across the equator. The fossil evidence is the same on both sides of the basin. Some shallow-water fossils have been found in the King's Trough region, a region now lying several thousand meters deep. Between the fossils and the continental rocks, one must pause to reflect on the magnanimity of these observations.

India, Australia, and Antarctica were never joined; Gondwanaland never existed. We have already seen that Sengor's models are globaloney, so Eurasia was never spread all over the Seven Seas either. So much for continental drift.

6-8. Mass Extinctions

Now we can have some fun. Perhaps you have heard about mass extinctions. They have happened many times over Earth's history. Mass extinctions are a normal occurrence in the evolution of life on Earth. Extinctions happened at the end of the Ordovician (440 Ma), Devonian (354 Ma), the Permian (245 Ma), the Triassic (202 Ma), and the Cretaceous (65 Ma), with many smaller ones to a total of twelve. In fact, we are in a mass extinction now.

Plate Tectonic Presumption: The Cretaceous-Tertiary mass extinction, and others, were caused by bolide impact, extraterrestrial bodies striking Earth. The K-T extinction was caused by such an event off the Yucatan peninsula.

Actuality: Carrying the geological misrepresentation into the popular literature, we are all familiar with the great dinosaur extinction at the Cretaceous-Tertiary boundary. Chicxulub crater in the Gulf of Mexico has been prominently featured in geoscience publications these past few years as the site of the asteroid (bolide) impact that caused the massive extinctions about 65 Ma. Charlie Officer and Jake Page made a more complete analysis of the rock material and refuted every one of the proofs, such as the ejecta blanket, DSDP Sites 536 and 540, and tsunami deposits (see *The Great Dinosaur Extinction Controversy*, Addison-Wesley Publishing Co., 1996).

In fact, Art Meyerhoff is the one who originally interpreted the stratigraphic sequence of Yucatan well #66 for Pemex in 1966. It penetrated an orderly sequence of Pliocene – Miocene, Oligocene, Eocene-Paleocene, and 350 m of Late Cretaceous sediments with Maastrichtian fauna above and Middle Campanian fauna below a volcanic sequence. No disturbance or hiatus exists at the K-T boundary, no asteroid impact was in this location, and these facts have been totally ignored by the mainstream ocean floor community. In fact, the core sample has now been "lost."

    Not to embellish the obvious, but we should by this time see that the name of Meyerhoff is anathema to any and all pseudo-scientists. It seems as though too many shares in the Brooklyn Bridge have been distributed! By the time of Art's death in 1995, he couldn't get anything published in his name any more. He had to use co-authors, which he did not mind. He would do anything to see the truth get out in print. In fact, even after his death, I have had comments related to Art's work stricken from my work. I actually had some idiot tell me at a GSA conference, when I questioned why their interpretation of the Pemex #66 site was wrong, tell me that he had recommended that GSA not publish Art's comments about Chicxulub. I had to wonder who had died and made him a god to know so much.

    The immediate question is: Do rock ages account for nothing in the field of tectonics? Is that hypothesis strictly a geophysicist's dream? Later articles have refuted Meyerhoff's ability to read rocks in the Pemex hole. His history speaks for itself; he would not have attained the lofty stature of being a world-class exploration geologist or head editor of the AAPG journals for over five years without some basic understanding of how to read rocks. All of his work in Southeast Asia and the USSR is still the standard, and the man has been dead since 1995.

    We have been bombarded these past few years by the liberal newsmedia as to the deleterious effects of our burning too much gas, the production of carbon dioxide, and the impending gloom-and-doom provided by anthropogenic greenhouse effect. Are we nervous yet? No. The greenhouse effect has been with Mother Gaea for many years, as in forever.

    Once the smoking gun was "determined" to be bolide impact, everyone hastened to change his/her respective views and jump on the bandwagon. Walter Alvarez won a Nobel Prize for his discovery of the iridium layer, iridium emplaced by an outer-space bolide. Things started happening.

    Then, someone took a look at the available stratigraphy. A great ice age is now thought to have caused the Ordovician die-off. The Devonian extinction was probably caused by the rise of trees, or even large tsunamis. This event was spread over 20 million years, so a bolide was certainly not responsible for this mass extinction. Along came the Permian, and with it the greatest die-off in history. Ninety percent of the oceanic species went belly up, and 70% of the terrestrial species joined them. The cause? The formation of the Siberian Traps with the attendant rapid climate change, lowering of sealevel, and poisonous acid rain caused by the magma's rise through three miles of coal. Additionally, the rise in temperature caused the oceans to warm, and this released vast amounts of pent-up methane, which raised the global temperatures even higher. A bolide impact is credited for the Late Triassic event, which gave rise to the dinosaurs. However, the evidence has been found in only a few places. School is still out on this one. The famous event at the Cretaceous-Tertiary (K-T) boundary about 65 Ma supposedly removed most of the dinosaur population thanks to the bolide off the Yucatan mentioned above. Thirteen genera went in, and nine came out to enjoy life in the Paleocene, including the modern-day theropod dinosaurs, the birds. Not much of a mass extinction, is it? Also, the stratigraphic record in Pemex Hole #66 does not show any signs of a disturbance at that boundary, so once again we have a fictitious hypothesis based on castles in the air. (Is there no pride in these "scientists?") The agent of destruction here has now been determined to be more probably the formation of the Deccan Traps in India. The last big mass extinction is occurring now, as many species are rapidly being lost. The Ice Ages are being blamed for this one, along with the rise of man, the hunter.

6-9. Who's Messing with Gaea's Thermostat?

Much to the dismay of the greenhousers, some folks have actually had the nerve to look at the real data on climate and its ill effects on mankind. The results are certainly not what was expected, and it will probably cause the tree-huggers to seek umbrage elsewhere. Oh, it's true, we are getting warmer, but let's look at the what, why, and wherefore.

The trends in weather and climate extremes do not show deleterious increases over time, instead the problem is that more people are moving into hazardous regions, and they are wealthier now than ever. This causes the headlines to scream about the foul weather, and the wealth makes us more vulnerable to the extremes.

I took a class in an MS I was working on during the early 1970s. One of the topics we discussed was the Outer Banks of North Carolina. The sand is constantly shifting, in this case southward. So, all the smart people who built beach houses out there would occasionally lose their investments. Naturally, the rest of us paid for this bit of foolishness by paying out increases insurance premiums. Well, the beach owners started putting up barriers to this erosion. This was great for the immediate area right in front of the breakwater. However, the prevailing currents eroded the heck out of the area just to the south of those barriers. So, that person also had to erect one. You get the picture, It's a never ending cycle, trying to tame nature. I think someone finally got the message and stopped development on sand dunes.

One study shows fewer low-pressure cyclones during El Ninos, and another study shows twice the dollar amount of damages during the 22 La Nina events since the 1920s as opposed to the 22 El Ninos. Additionally, the hurricane winds are milder during the El Ninos.

A Walsh and Pittock study (1998) revealed that, just like earthquakes and volcanoes, the climate models are unable to predict the effects of global warming on cyclones as (1) little relationship exists between the sea surface temperature (SST) and cyclone numbers, (2) there is little evidence that changes in SST could cause an increase, (3) there is little evidence that changes in ENSO could cause same, and (4) under increased greenhouse conditions, the number and severity of storms is greatly reduced.

Likewise, flood damage is the result of the conditions; more people with more money are moving into flood-plains. An analysis of more than 1500 stream-flow gauges throughout the continental US shows that the country is getting wetter, but that it is less extreme. In fact, the droughts are less severe than in the past 2000 years. With climate getting wetter, the farmers, livestock, and people in general are making out like fat rabbits, and there are less temperature extremes. What could be better?

Robert Felix wrote, in 1999, in *Not by Fire but by Ice* that 99% of the glaciers in the world are growing. One in Norway is inching forward, literally, about 7 inches per day. This is not exactly cause for alarm in increased temperatures.

I can tell you from personal experience of living in lower Mississippi from 1977-2000 that this study is true. We had less hurricanes according to the old-timers, and they certainly were not very fierce. The tornadoes caused much more damage. I only watched about three or four hurricanes the entire time I lived there.

The Greenland and Antarctic ice cores show climate fluctuations; they also show the CO_2 fluctuations for the past 110,000 years or so (see *The Ice Chronicles* by Mayewski and White, University of New Hampshire Press, 2002). I seriously doubt that the greenhouse effect was being perpetrated by anything prehistoric man was doing. In addition to carbon dioxide, methane gas (CH_4), nitrous oxide (N_2O), and ozone are the primary greenhouse gases. The present problems, believe this or not, are caused by two major contributors, termites and cows! The destruction of the equatorial forests and resulting slash left to rot on the ground has provided a feast for termites. They produce methane gas as a byproduct. Similarly, we now have more cows than ever in the past. Flatulent cows are overloading the CH_4 zone with gas. This is a true story. So, we are not headed for a mass extinction due to our excessive use of petroleum products. From this we may deduce that the possibility looms rather large that cows and termites may cause the death of us yet, excluding the "mad cow" disease.

Story: My last cruise before retiring was on the newly built SUMNER in 1997. I got a chance to go out on the new systems – type in the survey pattern at the start of the cruise and sit back and monitor about 10 screens! Talk about making life easier. Of course, any unforeseen course corrections threw everything out from that time. Anyhow, we were going into the South China Sea from Singapore. We had a constant supply of smog surrounding us, even up off Vietnam. I looked at one of our weather charts that came in every so often. It was a giant-sized plume of smoke from Indonesia. Seems as though the sugar "barons" were burning their cane fields in preparation for the new crop. I never did find out if the fires got out of bounds into the surrounding forest, but the air was plenty foul for about a week with an excess of rain. The plume showed up on the satellite maps and made the news I discovered when I got back home. In fact, Earth magazine called it "one of the most widespread man-made disasters ever." You worried about "greenhouse"effect? There it is, sistahs and braddahs.

It looks like stratigraphy, even ice core stratigraphy, and real climate data get the upper hand in this topic.

6-10. Deep and Superdeep Drilling Defies Geophysicists

Having never been on a drilling operation, I still feel the need to talk story. This was an interesting event that happened while we were on the search for the submarine, USS SCORPION. I was on the BOWDITCH. Rick Tyler, Herb Tappan, Charlie Beatty, and myself had the watch. We had had this movie, "Pinocchio in Outer Space," on board for about three months. It was horrible, and we kept shipping it off, only to get another copy back. Well, we were working with a task force the second cruise after we had found the missing boat; about a half dozen tincans, the MIZAR (a research vessel), and a Russian mother ship hovering in the background. We had positioned the transponder network for the MIZAR so that she could drag her photography sled over the wreckage. In order to get our chart-work to her, the higher-ups decided on a high-line transfer to one of the tincans. This was neat, because we hardly ever saw anyone on our SURVOPS, what with the nature of our work. Anyhow, while we were connected, about mid-cruise, our crew called the other ship about trading some movies. Of course, we all knew the first one we would put in the bag. The orange bags passed each other mid-stream. We noticed they were all out on deck laughing too. Damn if they didn't send us their copy of "Pinocchio!"

Plate Tectonic Presumption: The temperature increases with depth as one nears Earth's core in a predictable fashion, gradually rising to 8500°C. The rock types reflect the amount of heat at each interval, as rock density is generally expected to increase with depth and pressure. The expected sequence was 0-4.7 km of metamorphosed sedimentary and volcanic rock, a granitic layer from 4.7-7 km (the 'Conrad discontinuity'), and a basaltic layer below that. From the study of seismic waves, geophysicists have determined that between 7.5 and 8.6 kilometers below the surface there exists a clear-cut "discontinuity." Practically speaking, this means that above this layer seismic waves travel at a markedly different velocity than they do below it. This discontinuity is so widespread, occurring beneath all of the continents, that it has received a special name: the Conrad Discontinuity.

Actuality: Aside from the minor depths reached by spelunking, we now have some concrete data. Boreholes have been sunk into Eurasia. The deepest borehole drilled for scientific purposes is located on the Kola Peninsula near Murmansk, Russia, in the northwestern part of the Baltic Shield. The drilling of the main borehole began in 1970, and a final depth of 12,262 meters (about seven miles) was reached in 1994. A comprehensive report was placed on the Internet by the Russian Academy of Sciences, Geophysical Committee, which summarized the work done at this and several other deep boreholes.

Essentially, deep and superdeep borehole data has totally upset the geological applecart, so much so, that the Earth scientists who have made all of the predictions/presumptions now have a goodly dozen of eggs in their collective faces. One scientist commented: 'Every time we drill a hole we find the unexpected. That's exciting, but disturbing.' And a science reporter remarked: 'Kola revealed how far from truth scientific theory can roam.'

The densities did initially increase with depth. Then, at 4.5-km, the density decreased, presumably due to increased porosity. The results also showed that increases in seismic velocity are not necessarily dependent upon rock density. According to the Russian Minister of Geology: "with increasing depth in the Kola hole, the expected increase in rock densities was therefore not recorded. Neither was any increase in the speed of seismic waves nor any other changes in the physical properties of the rocks detected. Thus the traditional idea that geological data obtained from the surface can be directly correlated with geological materials in the deep crust must be reexamined."

Next, granite did not appear until 6.8 km, and it continued to the bottom of the borehole. No basaltic layer was found. The Conrad

Discontinuity beneath all the continents has been detected by seismic-reflection surveys heretofore, and it has heretofore been grossly misinterpreted. This was a shocker. Now, no one knows what the Conrad Discontinuity represents. It doesn't signal a change in rock type; neither is there a fault or boundary of any kind. It is important to find out what is wrong here, because much of modeling of the unseen structure of the earth's crust depends upon a realistic interpretation of seismic records. No change exists in the real world from granitic to basaltic rocks.

Other items of interest from the deep boreholes are that metamorphism was found down to the depth of seven kilometers, which should come as no surprise. Metamorphism is caused by the action of heat and pressure on already existing rock. Interestingly, hydrogen, helium, methane, and other gases, together with strongly mineralized waters were found circulating throughout the Kola hole. This was unexpected, as fractures at that depth and 3000 bars pressure were not supposed to occur. In fact, the Oberpfäalz borehole produced hot fluids in open fractures at a depth of 3.4 km. The brine was rich in potassium and twice as salty as ocean water, and its origin is a mystery. (The presence of all these gases is in line with the EMST hypothesis which we will learn about directly.)

The increasing temperature with depth assumption that Earth's temperature is 1000°C at 800 km, 4800°C at the core-mantle boundary, and 6900°C at the center is erroneous also. On the one hand, we have caves getting cooler with depth. On the other hand, deep mines and oil drilling have given us increases in temperature. What is true? Superdeep drilling has shown that temperature increases with depth far more rapidly than was predicted, reaching 180°C rather than the expected 100°C at the 10-km depth. However, at about 7-km, it started to decline. This is a good thing, because we would be hopping around like ants on a griddle were this rate to continue! It is bad for the geophysicists, though, because the mantle would be molten below about 100 km. Seismic survey data say that the mantle is solid. Who's right?

So, ultimately the drilling results at the deep and superdeep boreholes revealed significant heterogeneity in rock composition and density, seismic velocities, and other properties. Overall, rock porosity and pressure increased with depth, while density decreased. This could be accounted for by an increase in interstitial space. Additionally, the seismic velocities showed no distinct trend.

So, after several years of study, the Russian Academy of Sciences has found that the correlation between heat conductivity and temperature gradient in various depth intervals is "inapplicable." This affects " . . . experimental studies of metamorphism, earthquake prediction, dehydration, solid phase transformation, deformation processes, and others . . . "

As a result of the foregone, most of the modeling for Earth's interior has been categorically wrong. You and I have spent a lot of money supporting these erroneous presumptions, and we are in the process of spending a lot more. The continental seismic surveys, the basis of our "understanding" of earth's interior, have been misinterpreted, which parlays into a totally wrong picture of interior events, composition, and other geological and geophysical aspects too abstruse to be entertained here. And, this is only skin deep, the first seven miles of crust. Imagine how this plays out several thousand miles down at the core!

I don't know what all that means, either, but it sure doesn't look like we've ben getting our money's worth from these plate tectonic presumptions based on geophysical/mathematical models, does it? How can we make accurate predictions for volcanism and earthquake caused by the Mother's bowel movements if we do not understand the composition thereof? We have bought more shares in the Brooklyn Bridge, folks!

6-11. Earthquakes Defy Physics

And so, we now leave the world of tangible, hands-on facts to delve into the esoteric world of geophysics. Stand by for where this leads:

First, a quick lesson on how earthquakes are measured as to magnitude. From the US Geological Survey (USGS) Internet site: The Richter Scale measures the magnitude of seismic waves from an earthquake, devised in 1935 by the American seismologist Charles F. Richter (1900-1985). The scale is logarithmic; that is, the amplitude of the waves increases by powers of ten in relation to the Richter magnitude numbers. The energy released in an earthquake can easily be approximated by an equation that includes this magnitude and the distance from the seismograph to the earthquake's epicenter. The Ricther scale is an open – no lower or upper limit – scale, but no earthquake greater than about 9 has been recorded. An earthquake whose magnitude is greater than 4.5 on this scale can cause damage to human-built structures provided they are close enough to the epicenter; severe earthquakes have magnitudes greater than 7. The famous San Francisco earthquake of 1906 was 7.8 on the Richter scale; the Alaskan earthquake of 1964 was 8.4; and the Loma Prieta quake of 1989 was 7.1. Like ripples formed when a pebble is dropped into water, earthquake waves travel outward in all directions, gradually losing energy, with the intensity of earth movement and ground damage generally decreasing at greater distances from the earthquake focus. In addition, the nature of the underlying rock or soil at a particular location affects ground movements.

In order to give a rating to the effects of an earthquake in a particular place, the Mercalli scale, developed by the Italian seismologist Giuseppe Mercalli, is often used. It measures the intensity, that is, the severity of an earthquake in terms of its effects on the structures and the inhabitants of an area, e.g., how much damage it causes to buildings and whether or not sleeping persons are awakened by it. The Mercalli scale is a closed one, ranging from 1 to 12, at which upper limit all structures collapse; total destruction.

Earthquakes with magnitude (M_w) of about 2.0 or less are usually called microearthquakes; they are not commonly felt by people and are generally recorded only on local seismographs. Events with magnitudes of about 4.5 or greater – there are several thousand such shocks annually – are strong enough to be recorded by sensitive seismographs all over the world. Great earthquakes, such as the 1964 Good Friday earthquake in Alaska, have magnitudes of 8.0 or higher. On the average, one earthquake of such size occurs somewhere in the world each year. The Richter Scale has no upper limit.

Plate Tectonic Presumption: Almost all earthquakes occur at the plate boundaries, with the larger ones occurring in the subduction zones on a dip angle reflective of the direction of the descending slab. Strike-slip zones are outlined by shallow earthquakes.

Actuality: Into this we need to mix the June 1994 Bolivian earthquake, which has thrown a monkey wrench into the earthquake/subduction relationship, big time. This deep earthquake, an M_w=8.3 at 636-km deep, supposedly extended across a 30 by 50 km plane. It cut horizontally across the slab and extended well beyond the supposed olivine layer. The significance of this fact is that, heretofore, the motion of earthquakes at depth had been expected to be nearly vertical. The horizontal movement of the Bolivian earthquake is difficult to explain within the current constraints. Big, deep earthquakes should not happen at all because "the enormous temperatures and pressures at such depths should allow rock to dissipate stress by flowing quietly rather than fracturing suddenly ... "

A prodigious amount of heat, on the hypothesized order of 35 billion megawatts, should not have happened at this depth. The pressure equivalent at this depth is about 200,000 atmospheres, so an excess amount of melting should occur. Hypothetically, this is the very bottom of the descending slab, a slab which begins the melting process many kilometers above this point as it feeds fresh magma back to the surface in the form of active arc volcanism. Where deep fault zones have been exposed on the surface, no evidence of any kind of heat or melting can be found, though.

The Shikotan Earthquake in Kamchatka presnts another anomaly. This 1994 M_w=9.3 quake occurred in an intraplate transverse zone in the Kuril-Kamchatka arc. It again demonstrated a "non-subductional mechanism."

Credit where it is due, the 1997 Kronotskoe earthquake is included. With an M_w=7.9, and an M_w=7.5, this quake defined the subduction plane as presumed. One out of three ain't too bad.

Then we have the horizontal earthquakes, as in strike-slip zones. The Landers earthquake, occurring 28 June 1992 in southern California, was a magnitude M_w 7.3 right-lateral strike-slip to the depth of 1.1 km. The average slip was about 3-4 m, with 6 m being the max. The accelerated amount of strain following this earthquake resulted in anomalous post-seismic displacement of the crust ranging from a few mm to 55 mm along the fault, resulting in the Big Bear aftershock, which registered M_s 6.4. The strain released in the upper 10 km of the lithosphere can be easily explained by aftershocks. However, the major release came from below 10 km, showing viscous relaxation along more than 100 km of the fault.

We have access to several earthquakes around eastern Kamchatka. One in particular, the 1995 Neftegorsk quake on Sakhalin Island, have never occurred in this region before. With an M_w=7.1, and an M_w=7.6, this was a hoss. There was a rupture length of 40 km, and the fault plane is nearly vertical. There was an absence of aftershocks, but the motion was as expected by the plate-tectonic interpretation for this region.

India represents a prime laboratory for the study of earthquakes not associated with convergent margins, having undergone several in the 1990s and early 2000s. The 1990 Uttarkashi quake measured M_w=6.6 at 10-15 km deep. The 1993 Latur quake recorded an M_w=6.2 at 3 km deep. Interestingly, this earthquake occurred in Archaen granite – gneiss basement of the craton, killing over 10,000 people in the process. Nasty. Then a relatively minor M_w=5.8 quake hit at Japalpur in 1997. It occurred just to the south of the Narmada-Son Lineament at 35 km deep. Last, the second worst in India's history hit the province of Gujarat in 2001. This M_d=7.9 monster leveled the city of Bhuj, killed over 20,000 folks, and injured 167,000 according to reports posted on the Internet. The depth was 23.6 km.

These earthquakes are not at the Himalayan border. They, in various arrays, are alined with others showing a linear belt from the Kashmir cusp southward through the Punjab and Rajasthan depression down along the western coast to join the juncture of the Chagos-Laccadive Ridge, which lies offshore. The features lying along this route get relatively younger from Pakistan through the Deccan Traps, a feature which lies at the intersection of these two trends.

The Mendocino Megatrend was long ago predicted to continue to at least Yellowstone National Park in NW Wyoming. We know that earthquakes are associated with megatrend activity. A report from the Utah seismic network (UUSS) indicates at least 13 events in the park from 10-17 October 2003 (as I am writing this booklet), with an additional 10 in the Teton Range. Mapping the events from 1979 to the present gives a clear outline from NW Yellowstone through the park to the ESE, turning south, and continuing SSW through the Tetons. The present swarm has M_s ranging from 0.4-1.9, all lying above the 350° isotherm in the brittle crust. Some of the roads have been shut down to tourism, and many of the birds have left the area. This is similar to tectonic events before the eruption of Mt. St. Helens in the early 1980s. And, Yellowstone is not anywhere near a plate boundary/subduction zone.

Last, the USGS has been studying earthquakes in New Mexico. Their data, collected between 1962-1998, shows no evidence at all of any activity along the state's major structure, the Rio Grande Rift.

Deep earthquakes can happen anywhere. They seem to be controlled by the deeper tectonic structures about which we can only hypothesize.

6-12. Mantle Plumes vs Sub-Lava Flow Stratigraphy

One of my more interesting cruises was with Don Hussong and Patty Fryer of the Hawaii Institute of Geophysics. We were doing a total survey of the Mariana Trough region when we came upon an island called Maug, which is Guam spelled backwards. Maug is surrounded by active volcanoes and is, in fact, three separate islands that form the crater's rim. Don convinced the ship's master to go inside the crater for a side-scan survey. Talk about awesome. Anyhow, after making the rounds, Patty decided to take a zodiac over to collect a few rocks. I have always wanted to do that but, alas, was not allowed to go. We did manage to dip and pour some of the water over ourselves so we could at least say we had been "on" Maug, a totally uninhabited island in the middle of the Marianas. Never did find out what happened to those rocks. As a matter of fact, I don't think they ever processed the survey data either.

Earthquake data can be used in other fields. Earthquakes have several different signatures, as in "P-waves" and "S-waves." The speed of sound an earthquake event makes through different media, as in solid, liquid, or water, is measurable. Theoretically, as the substrate becomes more dense, the speed increases. Conversely, as the substrate becomes *earthquakes — high heat flow areas*

more liquid, the speed decreases. Therefore, the speed of sound an earthquake makes when going through Earth's mantle should decrease. This is a simple connect. However, the speed increases the deeper the shock wave goes into the mantle, to a depth of about 2900 km! What this means is that Earth's mantle gets stiffer as depth increases, and not the opposite, as plate tectonics and the heat engine earth imply.

Seismotomography is another of the 25-cent words associated with tectonics. Seismotomography uses teleseismic data to give an image of Earth's interior. The velocities of the earthquake P – and S-waves passing through Earth's interior are measured at over 1000 stations worldwide. Digitized velocity information along each ray path is matched to corresponding velocities, giving the information needed to diagram Earth's interior in 3-D. By using this method, high – and low-velocity regions can be delineated. The high-velocity regions are assumed to be "cold." The low-velocity regions are assumed to be "warm."

Plate Tectonic Presumption: Large convection cells are one of the proposed mechanisms driving plate-tectonics. That hypothesis calls for a band of high heat flow, or low velocity, beneath the midocean ridges and landward of the trenches, with bands of low heat flow, or high-velocity, seaward of the trenches and all of the interior of the plates.

Actuality: What seismotomography has shown is quite the opposite. Heat flow studies in 1984 based on seismotomography established a belt of elevated heat flow. This belt is predominantly equatorially circumferential with cold areas east of the trenches, continental cratons, and generally midocean. Seismotomography has shown that organized convection cells do not exist in the mantle. By combining the seismotomographic and heat flow studies with micro-earthquakes; that is, earthquakes between 3.5 and 5.5 magnitude, the reticulated micro-earthquake bands correspond exactly with the regions of high heat flow.

Seismotomography has shown that the continental craton roots are deep, deeper than the maximum of 200 km needed to let the continents enjoy a free ride on the peripatetic plates. The depth of the cratons in places, i.e. North America, can be 5-600 km. In fact, the

in NA could be up to 500 km

↓ 200+ to allow continents to move — pillars of the earth

Continents have never moved

NA → 500 km deep

conclusion has been reached by some that the continents are fixed in position because of the root depths. They have never moved from their current positions.

Using earthquake seismology, the concept of convection cells as a motivator of plate tectonics does not work. Simply put, how can a solid, 7 km thick crust coupled on 100 km lithosphere moving around on a surface that is not exactly a spheroid push a 600 km deep continent? Seismotomography, even in its earliest form during the mid-1980s, showed that, rather than a hot region seaward of the trenches, a cold zone exists. Later refinement has given a pattern of heated, reticulated bands covering the globe, almost all aligned in an east-west direction. Most of the tectonic belts lie along one of these bands. This will prove to be significant.

East-West

6-13. Last of the Red Hot Mamas: EMST and a Cool Core

In 2001 I was invited to attend a workshop/symposium sponsored by the New Concepts in Global Tectonics Working Group. At the last minute, the Russian convener of the conference got a Fulbright scholarship to Otero Junior College in Colorado, so we all went there. On a field trip through the southern Colorado Rocky Mountains, hosted by the college, we were shown many dike structures, especially in the region of Spanish Peaks. We would expect to see this around volcanoes, but this case presented some very unusual facts. None of the country rock was disfigured/displaced/ metamorphosed due to the later emplacement of heated material. In geologist terms, no "baked contacts" were present. It was as though the country rock had just moved apart and allowed another rock to replace it in space, done in a cold environment. This same phenomenon occurs in the Alps and in eastern Australia.

Cold environment

Plate tectonic Presumption: When a body cools, it contracts in most physical cases. Earth is theorized to be such a body that cooled until at least 200 Ma. The amount of contraction has been to be no more than 2-3%. Additional evidence points to a cooling process that may have

[Handwritten at top: Core + mantle — cold. Crust — high temperatures]

ceased altogether by 200 Ma. The tectonic processes are related to a heat engine Earth.

Actuality: Harken unto thee; Stavros Tassos of the Greek Seismologic Institute in Athens has actually used the earthquake information and come to a startling conclusion: Earth's interior, namely its mantle and core, is cold, and only in the crust can high temperatures, and therefore melting, locally occur! Stavros was standing right there when we saw those cold dykes in Colorado. Not being a rock-hound, I had to ask why there was no deformation around those features. This is his explanation, edited by him: *[Handwritten: Not heat]*

Earth geodynamics are most ~~likely the result of~~ quantum-mechanical, and not heat related, processes. As we have seen, the geophysicists have given us close to the melting point of middle Earth, based presumably on the planet's formative thermally controlled stages. The Kola drillsite has given us increasing temperatures at least to 12.5 km deep, so this is concrete proof, of local high surface temperature. But, high temperatures can only be a spatially and temporally limited surface phenomenon, since the estimated global production of magma and lava is between 4 and 20 km³ per year, an absolutely insignificant amount as compared to the $\sim 10^{12}$ km³ the total volume of the earth, or even the $\sim 10^{10}$ km³ volume of the crust, and when we go deeper earthquake seismology tells us something else entirely.

[Handwritten margin: cold dykes]

The earthquake velocity through a heated, plastic to liquid body, such as we are being told exists in nature, should slow down. It does not; the velocity increases as it passes through Earth's mantle, and attenuation, which is 1 over the Q factor, in the outer core is the smallest possible, since Q is equal to about 10,000, as compared for example to $Q = \sim 100$ for the crust. This means that the core is liquid all right (since S waves do not transmit), but a cold, almost friction free liquid, in a high energy/high frequency – higher than 10^{14} Hz – environment. If the outer core was made of molten iron the Q factor would have been much lower, and therefore the attenuation much higher. Additionally, the sum of energy needs of the earth on an annual basis is of the order of 6×10^{14} Watts. If Earth is indeed a heat engine, there must be the equivalent heat source,

[Handwritten: heat source?]

which every year can provide that amount of heat. There are three supposed heat sources: 1) primordial heat, that is trapped heat at the time of Earth's birth about 4.6 billion years ago, 2) radioactive elements, and 3) tidal heat. The supposed 6×10^{22} Watts of primordial heat could have only lasted for the first 100 million years, with the present annual energy requirements. Radioactive isotopes, such as thorium 232, uranium 238, and potassium 40, are known to exist only in the upper layers of continental crust rocks, and in concentrations that do not exceed 21 parts per million (ppm). The heat equivalent is only 3×10^8 Watts, that is the 0.0000005 (!) of Earth's annual energy needs. In other words if all the energy needs are to be supplied by radiogenic sources their concentration should be 42,000,000 ppm or an earth 42 times its present quantity of rocks, and all made off of radioactive elements!!! As to the 10^{12} Watt of tidal heat, which is due to the about 11 cm uplift of the mantle due to moon's gravitational attraction, is sufficient for only the 0.0016 of its annual energy needs. Based on all of the above, Tassos has determined that earth's mantle and core are cold, and geodynamic processes are not thermally driven. This, of course, negates all previous tectonic hypotheses, based on a heat engine Earth.

Not satisfied to leave it there, Tassos has been working on the cold interior problem for some time now. The introduction of the iron-rich 'excess mass' increases internal pressure, forms micro-cracks, initially of atomic size (10^{10} m), 'new' electrons from the Fe^2 that now turn into $Fe^{2,3+}$, are discharged into, further increase internal pressure and widen the micro-cracks to crystal size (10^6 m). Simultaneously these 'new' electrons in the 10^6 m microcracks, can resonate with the 'old' electrons of the metallic bond at the 10^{14} Hz thermal-infrared radiation frequencies, thus producing at first heat, and finally the hammer blow, the adiabatic deformation and the earthquake. Instead of gravity mechanics, Earth is thought to be governed by quantum mechanics. The lack of transmission of S waves in the outer core and its extremely low dumping (high Q factor) is taken to imply a friction-free, superfluid core. These characteristics do not favor a hot iron-core model; rather these characteristics favor large amounts of helium in the outer core. Along with its dielectric property, the presence of helium can also explain magnetic reversals.

Called "Excess Mass Stress Tectonics," or EMST, particles-plasma constantly transform into atoms. Iron is the last atom to form, because it is the atom with the highest nuclear binding energy, i.e., 8.8 MeV per nucleon. The transformation takes place in the outer core. The newly formed atoms are added, one-by-one, as solid wedges in the preexisting and overlying mantle and crust. The resultant is the density of plasma and, therefore, of the outer 'cold liquid' core to be reduced. This is not necessarily its volume, and the density and the volume of the solid mantle and crust will increase. This process will continue until all the plasma of the core is transformed into atoms that are being added in the mantle and the crust and generate all geodynamic phenomena, including earthquakes and volcanoes. After the completion of this transformation, Earth will become a magnetically and tectonically inactive planet, like the Moon and Mars are. The basic premise of the EMST concept is the transformation of simpler, smaller hydrogen and helium units into bigger, more complex ones through electromagnetic confinement, laser clustering, and nuclear fusion. (We have seen that an excess of hydrogen and helium has been found as close to the surface as 12-km in the superdeep boreholes.) Two phases of E.M. generation are recognized. The first time was from 4.0 Ga to 0.2 Ga, when only a small fraction, ~3%, of the initial number of nucleons was transformed into iron-poor, but Na, K, and Ca rich, continental crust rocks. Besides Precambrian shields, the surfaces of the Moon, Mars, and Mercury (still passing through the first stage), are cited as examples. The second time was from 200 Ma until the present, when about 65% of the nucleons have been transformed into iron-rich mantle and oceanic crust rocks.

The core of the Earth is an electrically unbalanced gas of particles which are subject to the exclusion principle. The degeneracy pressure due to electrons should be greater than the gravitational pressure due to nucleons. The degeneracy pressure is reduced during periods of electron clustering. During periods of intensified clustering, the degeneracy pressure is reduced, and the Earth contracts somewhat. The net result of the electrical imbalance is the pulsation of the Earth, which is superimposed on its prevailing expansion.

The EMST hypothesis attributes phase changes to the upward

movement of E.M. (not as hot magma) from a high pressure/high frequency-low temperature deep environment to a low pressure/low frequency-high temperature surface environment, i.e., from perovskite to spinel to olivine to eclogite to basalt, and the phase changes have nothing to do with the earthquakes. Faults and earthquakes are interrelated but different in their generation mechanism phenomena. Faults require very weak stress, e.g., inertial stress, due to weakening of the rock and the unification of micro-cracks, which are caused by an increase of internal pressure, while earthquakes in order to occur a high stress rate of the order of 10^{15} Pa.sec^{-1} is required, many orders of magnitude greater than the 10^{-2} Pa.sec^{-1} thermal-gravity generated stress.

In a hypothesis called oceanization, an all-encompassing thick continental crust covered the smaller earth, which essentially made the planet one large continent. It had on it narrow and shallow seas. Below, the continental crust that was also formed by solid intrusions was not the iron rich mantle, but the cold plasma outer core. That is, electrically charged particles, which with the process of self-organization, transform into atoms. Thus earth's interior operated and continues to operate as a natural laboratory producing new elements added as old 'solid wedges' into the superjacent material.

The newly formed material, the 'excess mass' relative to the mantle and the crust, like a nail, creates micro-cracks and fills them up. That way a reasonable explanation is offered to the observation that seismic wave velocity is inversely proportional to density; the introduction of the nail increases the density but at the same time creates micro-cracks that break the continuity of the medium, and thus lower its elasticity and consequently increase attenuation and lower the velocity of seismic waves. The main component of the 'excess mass' during the last 200 million years is iron. Its introduction results in richer in iron minerals and rocks, and this is oceanization. Gradually as more and more 'excess mass' is being added the splitting of the old continental crust occurs, the gap is filled with the iron-rich mantle and oceanic crust, and finally an ocean basin is formed.

The EMST explanation calls for cold transfer from inner Earth. However, we all know that magma/lava exists, so we must derive an explanation for that phenomenon. The basic assumption from a heated interior Earth is that the magma rises from the inner sphere to fill the asthenosphere with molten rock. From a cold inner Earth, radiation at about 10^{14} Hz; that is, the thermal-infrared vibration that can be achieved in 10^6 m micro-cracks. The micro-cracks remain open at depths above 30 km in the lithosphere. Because the hypothesized hot lines must be fed, we now assume that the micro-cracks are more-or-less aligned with the macro-cracks (fracture zone/megatrend), and the radiation in the micro-cracks feeds magma into this system.

Giovanni Gregori has also experimented with the cold fusion-helium link along with joule heating. John Quinn, late of NAVOCEANO and the US Geological Survey, has used magnetics to determine the same thing. In an interesting experiment with the electrical properties of Earth, Karl-Heinz Jacob of the Technical University of Berlin noticed desiccation in house walls. He found that the electrical potential differences in nature are great enough to transform minerals from one state to another. By sending electricity through water, he was able to create thin bands similar to secondary structures of Leisegang rings, undulatory and fold-like diagenetic banding, micro-diapirs, "boudinage" or "breccia," rosettes of varying dimensions, and fractional structures, such as dendrites and cauliflower structures. The upshot is that "nature really works via self-organization."

From these facts we deduce that Earth is a cooled body rather than a heat engine, and that cold fusion is one of the driving forces of tectonics. Hence, all previous ideas as to the driving forces of any kind of tectonics are off, at least in relation to heat supplied by core and/or mantle processes.

By this method the megatrends are refilled with magma on a periodic basis. If there is an asthenosphere, it is not everywhere the same thickness. Channels may form both above and below the asthenosphere proper. Because of the existence of so much earthquake information above 150 km, and the spotty occurrences of deeper earthquakes, we feel that the earthquakes outline the hot lines. The hot lines are the channels, and they are defined by systems of micro- and macro-cracks and linear seamount chains.

From this information we may add a cold, liquid outer core of a possibly expanding Earth.

6-14. Fat Bottom Girls, or Gaea Does Not Have a Middle-age Spread

Realizing that something was amiss with the plate tectonic hypothesis, in 1976 Sam Carey reopened the case for an expanding Earth. The models proliferated, several using nothing more than Earth expansion based on seafloor spreading/ocean floor magnetics with no subduction. This limited the models to conditions from the present back through the Jurassic because, as noted, no magnetic anomalies exist that are older than 180 Ma. Carey, Klaus Vogel, Oakley Shields, and James Maxlow all have a fast-expanding Earth that was only 60% of its present size in the Jurassic. Carey believed the equatorial region to be in a state of sinestral torsion, which was thought to be the combined effect of gravity and rotational inertia. Carol Strutinski changed that to mean an easterly flowing asthenosphere, and that this could also be inferred from Jupiter and Saturn.

Vogel's global models provided what some considered the most convincing model of Earth expansion. He had most of Earth covered by continental crust until the Mesozoic, with the recent crustal movements all being related to radial outward pressing of the continents and the infilling by new oceanic crust from ocean floor spreading at the midocean ridges. Shields used paleobiogeography: "In Permo-Triassic times, e.g., there are terrestrial biotic links spanning the eastern Tethys and central Panthalassic oceans that don't occur across Pangaea, so explaining these seems to require an expanding Earth." G.O.W. Kremp hypothesized that Earth was only 40% of its present size at 2.5 Ga. H.G. Owen is a little more conservative in that his Earth was already 80% of the modern size during this period of "great expansion." Owen used the same geometry exercise used herein to prove that the generation of crustal material does not equal to the amount of subduction plus the amount of foreshortening (collision). Michihei Hoshino felt that Earth has expanded very little, but that it is expanding based on glaciology and sealevel changes without the agent of crustal subsidence.

All of these models include descriptions of the growth of the Tethys Sea and the fit of Pangaea, presumably based on magnetic measurements produced by a heat engine Earth. They all include at least three phases of Earth evolution. The continental shield/granite stage was during the Archaean Era. The original continental crust was created by calc-alkalic magmatism before the Mesozoic in a series of steps. These in order were geosynclinal, orogenic platform, continental rift, and block tectono-magmatic activization. The Proterozoic and Paleozoic Eras were called the transitional stage. The Mesozoic and Cenozoic are called the basaltic stage. Basalt is replacing granite from the midocean ridges by seafloor spreading.

In order to prove Earth expansion, the moment of inertia constraints must be overcome. An expanding Earth would necessarily rotate slower than a smaller diameter planet. This is called "conservation of angular momentum." In order for this to happen, the lunar tides would have to slow down, which would affect the length of the lunar month. Here the expansionists rallied for many hypotheses. A study of the growth patterns of mollusk shells since the Ordovician shows an Earth year of 447 days at 1.9 Ga decreasing to an Earth year of 383 days at 290 Ma to 365 days at this time. Another parameter, the Devonian coral rings, shows that the day is increasing by 24 seconds every million years, which would allow for an expansion rate of about 0.5% for the past 4.5 Ga. According to the coral growth patterns, this is not enough to warrant a discussion, especially in terms of ocean basin opening, moving continents around, or the dispersal of Pangaia.

Actually, during the past 900 Ma Earth has experienced a slowdown in spin rate according to the NASA space physicists. Even now, this can be detected, and the rotation rate changes in milliseconds per day. This is dependent upon "how the mass distribution of Earth and its atmosphere change from earthquake and the movement of water and air." A further explanation of the spin slowdown reveals that a day was 18 hours long 900 million years ago. Because the Moon and Sun are constantly applying a tidal force to the Earth, the water moves as tides. As that water flows against the ocean floors, it applies a "braking force," essentially a slow-down mechanism, on the planet. In billions of years, this will force the lunar months to increase from the present 27.3 days to 47 days.

Chandler wobble and axis tilt also change the rate-of-spin. If the tilt were as much as 54° instead of the current 23.5°, the polar regions would have had the regional warm climates that have produced the fossils being found there with no continental movement. Also, the motion of large air masses can change due to tilt by measurable amounts daily. The change in tilt affects the rate of rotation. Chandler wobble "resides in the natural resonances in the body of the spinning earth due to detailed distribution of mass in its surface, interior, oceans, and atmosphere."

Primarily, though, Earth expansion is a discussion of philosophies, and this philosophy is based on math, statistics, geophysics, and theory. Real data, such as rocks and bathymetry, provide the best evidence. Words seen in researching expansion were "estimates . . . should be . . . postulated . . . " In simplified terms, Earth expansion is plate tectonic seafloor spreading without subduction during the past 200 Ma.

Earth rotation and easterly flow are added to our shopping cart from this section, but school is still out on Earth expansion.

7

So, What Do We Have Here?

THIS STORY CONCERNS the very work we did at NAVOCEANO. It seemed to us that many of the academicians had ocean floor features named after them, some more than several. Menard and Heezen had about 3-4 each. When I looked into the naming procedure by the US Government Board on Geographic Names, I found that the discoverers usually could name the feature being discovered whatever they wanted to. Good and well. I named features after our ships (such as Michelson and Dutton Ridges), after characteristics of the feature (such as Broken-Top Guyot), or even took the liberty of recognized some of NAVOCAENO's most senior seagoers (such as Manken, Jensen, Musgrove, Beatty, and Vibelius Seamounts). I used the criterion of having taken over 100 voyages of discovery, certainly no slack feat by anybody's standards. Well, you guessed, one of the keepers of the keys (from Scripps) protested in an Opinion article in Geology. Naturally, I took offense at this backstabbing and looked into his credentials. He had a Fisher Ridge and a Fisher Seamount already named after him, and I'm quite sure that he hasn't been to sea half as many times as any of those I used. Fortunately, the USBGN was not influenced by his petty attempt to discredit the Navy surveyors, or me. I also received a letter from yet another academician (from Lamont) about my "hubris" at naming features. Not to worry either. Considering the amount of globaloney they have imparted to us over the years, at a near-total waste of taxpayer dollars, I feel no remorse whatsoever about anything I ever did.

Thomas Kuhn wrote about the structure of scientific revolution some years ago, and his idea has been updated. A paradigm can only work if it explains all questions, and this aspect is called "Occam's Razor." This occurs when the facts gathered overwhelm the existing paradigm. However, the old will not give way until a new paradigm is in place to replace it; something that covers more of the bases. This is defined as a "scientific revolution." Max Planck figured that most of the older generation who derived the old paradigm must die off before this could happen, making an overturn more likely to occur about 35-40 years after the formulation of the current working paradigm. Essentially, Kuhn says that the keepers of the key will do anything to keep any outside influences from penetrating their inner sanctum. The failure of the existing plate tectonic paradigm opens the door for something new.

The primary working hypotheses, as we have seen, were formulated in 1966 and 1976, more-or-less. Oh, they have been tweaked, fine-tuned, and re-calibrated since that time, but they are basically the same as they were upon origination. What I will do here is to analyze the principle facets of both plate and expansion models, see what, if any, new data has been found to support/refute the original tenets, and apply the new data to a more robust working hypothesis if necessary. This time, in other words, we will try to fit the hypothesis to the data rather than the other way around. After all, Occam will have his day, and plate and expansion tectonics should certainly be able to withstand close scrutiny.

What? Flies in the ointment? And we have barely scratched the surface of the available data. This treatise could go on ad infinitum, ad nauseum . . . This is going to get very interesting, and more than a few "scientists" are going to get caught with their collective hands in the proverbial cookie jar.

First, most of the planets in this solar system are already cool. Their tectonic cycles have been fulfilled. No reason exists to believe the Earth is not also already cool. Second, Gaea is always rotating while revolving around the sun, and that creates centrifugal force. This phenomenon is easily demonstrated with a bowl of water. As you slowly turn the bowl, the water swirls in the same direction. It does not move as fast, but it moves in the same direction. Some drag exists to hinder the progress of the outside edges of the water. Third, we have all heard about gravity. Hold this book in the air and drop it. A man named Newton discovered that years ago. So, on a planetary scale, we know that we have centrifugal force and gravity acting on Earth.

A recapitulation gives us much to work with. We have (1) a geometry that won't work, (2) mis-interpreted magnetic anomaly data, (3) along-ridge magma flow, (4) rocks in excess of one billion years old on the midocean ridges, which is essentially supposed to be brand new crust, and (5) interconnected heated channels. Those problems were all solved in the section on midocean ridges. A big problem that was listed was the origin of the gravity lineations in the central Pacific basin. The Central Pacific Megatrend answered that. We have discovered (1) fracture zone intersections, (2) seamount chains associated with fracture zones, (3) vortex structures, (4) horizontal earthquakes, (5) a pre-Carboniferous Tethys Sea, and only one of them, (6) India having never been a part of Gondwanaland, (7) nothing in the rock sequence to show a bolide strike off the Yucatan peninsula, (8) no subduction, (9) Earth rotation, and (10) easterly flow in channels. Sounds like the fixin's of a good pot of gumbo.

For those who think we know it all: If P-T conditions of the Earth were what is being taught and accepted as established, there shouldn't have been very deep earthquakes like the one of June 1994 in Bolivia, which had a magnitude of 8.3 and cut horizontally across a 30 x 50 km plane at 636-km; or, we wouldn't have encountered highly fractured rocks at 12-km depth during drilling in Kola Peninsula. Up on the Earth's surface, the gravity lineation of the Central Pacific Megatrend doesn't fall in any plate tectonic setting.

Notwithstanding the fact that plate tectonics is accepted and acclaimed as the working geodynamic model, the fact is that the basic elements of the model itself remain un-established even after almost four decades of exploration. But this comes as no startling revelation to any serious student of ocean floor science. This is evident from the type of problems being proposed for study now. For instance, of the several unresolved problems of plate tectonics identified for investigation under the Ocean Mantle Dynamic Initiative (home page, 2000) the following three tell the whole story: (a) "along [ocean ridge] axis mantle flow", (b) whether interconnecting "channels" exist between ocean ridges and off-axis plumes, and (c) the origin of gravity lineation in Central Pacific Ocean, which defy any and every conceivable plate model. Along axis mantle flow is at cross purposes with the orthogonal mantle flow, the building block of plate tectonics model!! Add to these the unresolved problem of "driving mechanism" (also identified for investigation under OMD), what is then, or what is left of, plate tectonics!! Relevant to this study, the worst failure of plate tectonics as a geodynamic model, irrespective of its so-called "all-embracing character", are: a) its inability to include ocean and atmospheric dynamics within the overall Earth dynamics; b) its treatment of land and sea distribution – so critical to climate – as mere coincidental (rather like pieces of drawing/living room furniture, shifting places depending upon the mood of resident/plate tectonics modeler); and c) its inability to find a place for rotational force – again, so basic to climate – in the overall framework of the model. Folks, this is getting serious. None of the basic premises holds water – not one. It is obviously time for a paradigm shift.

In the September 2003 issue of the New Concepts in Global Tectonics Newsletter, the editors refer to recent discoveries concerning Earth's fundamental facts. "These include the increasing relief of the earth (Ming Xu Gao, Cliff Ollier, and Colin Pain). Is this connected with the formation of the present oceans and continents from the beginning of the Jurassic (Mac Dickins, Dong Choi, Tony Yeates, Igor Rezanov, Boris Vasil'yev, and others), the hypsographic bimodality of the oceans and continents apparently related to the Jurassic and lower Cretaceous events and formation of the continental shelves apparently from the mid-Cretaceous? Is this connected with the pulsation of the earth with expansion of the crust and foundering from the Jurassic to mid-Cretaceous and contraction and increase of relief of the earth from mid-Cretaceous to the present with analogous earlier cycles (Mitsuo Hunahashi, Colin Laing, Forese-Carlo Wezel, and Dickins), the relationship between deep structures and upheaving and the position of geosynclines (Hunahishi, Yukinori Fujita, and Yasumoto Sujuki), and the lineaments (Smoot) and deep earthquakes (Choi)?"

An old established scientific dictum states that one recalcitrant fact is enough to wreck a hypothesis. A list of 29 geological and geophysical data sets not only remains unexplained by plate tectonics, but conclusively negates its viability as a geodynamic model. This list was compiled at the 1988 International Geological Congress in Washington DC. It has been published in several media, and all of the questions remain unanswered/ignored 15 years later. Meanwhile, our tax dollars have been supporting this "research." As most of the topics have been covered, or are about to be discussed, a list is not necessary. Memorizing long lists is too much of a hassle, so we won't proliferate that bad boy.

8

Gaia's Basic Forces: An Updated Working Hypothesis

I HAD ALWAYS tried to recognize and give credit where credit was due. When we finished our surveys of the Emperor Seamounts, I wrote a couple of papers; nothing too exciting, mind you, just to show what they really looked like. I got a response from an interested, and apparently irate, reader who claimed to have discovered a portion of the seamounts and named them. So, I went and found his paper. Then I told him that he had two seamounts where there are really five. I told him to pick out which two he wanted, and I would take the rest. Never heard from him again!

Luckily, the cooks have already been in the kitchen and derived a working hypothesis. It may or may not be the final answer, but it does incorporate all of the facts we have collected in this treatise, and it is called "surge tectonics." Surge tectonics is a newer geodynamic model which has risen out of the geological and geophysical data gathered over the last century, particularly during the later half of it. In summary, the model is based on the concept that the lithosphere contains a worldwide network of deformable magma channels (surge channels or geostreams) in which partial melt rises at discrete locations from the asthenosphere and is in motion parallel to the trend of the surge channel. Surge channels could be active if the magma is still in motion or inactive if the magma was in motion some time in the past. They occur at depths between 50 km and the asthenosphere and underlie all conceivable small and large scale tectonic elements: ocean ridges, aseismic ridges, mountain belts, rift zones, strike-slip zones, etc. All active surge channels are characterized by higher than normal heat flow (>55 mV/m^2), tectonic element-parallel faults-fractures-fissures, bands of microearthquakes, thermal springs, elevated groundwater temperatures and volcanic phenomenon. The fault-fracture-fissure systems are interpreted to be streamlines, indicating that the mechanism producing these features is viscous drag resulting from fluid motion parallel (*not* orthogonal as in plate tectonics) to the trend of the tectonic feature. The presence of surge channels means that all of the compressive stress in the lithosphere is oriented at right angles to their walls. The mantle above the strictosphere (i.e., hard mantle below the asthenosphere) resembles a giant hydraulic press. A hydraulic press consists of a containment vessel, fluid in the vessel, and a trigger mechanism. In the case of Earth, the containment vessel is the interconnected surge-channel system; the fluid is magma in the channels; and the trigger mechanism is worldwide lithosphere collapse into the asthenosphere when the latter becomes too weak to sustain the lithosphere dynamically, causing the magma in the channels to surge out intensely. The ultimate cause for all motions in the asthenosphere and the lithospheric surge channels is Earth's rotation, which explains many of the tectonic features e.g., the presence of island arcs (e.g., Japan and Philippine island arcs) in the east of Asia but not west of the Americas, chain of ocean islands originating at a cusp where two island arc intersect, etc.

That channels exist within Earth's lithosphere is amply demonstrated with the section on MORs. They lie in the 50-150 km deep range based on earthquake data; or, in this case, the lack thereof. Rather than a voluminous constant outpouring, most of the magma remains in the channels, or hot lines. Those of us who live near active volcanoes may not think so, but this is a fact. Iceland itself can have about 75 km^3/yr added to the surface, and it sits atop the Mid-Atlantic Ridge. One blast from Mt. Etna on 27 October 2002 produced 10-11 km^3 of lava and more than 20 km^3 of tephra, and who knows what kind of geologic feature it sits upon. Interestingly, the USGS Internet site "clearly" states that this figure averages 3-4 km^3/yr! I live in Hawaii; Kiluaea alone seems to produce that much.

If this much is true then, because of Earth's easterly rotation pattern, the magma goes through the channels in an easterly fashion because that is the direction of the force field. It will not always be due east; it will find the path of least resistance, just as we do, and follow that lead. Where it cannot break out to go easterly, it will pond, eventually building an eastward bulge. The western Pacific forearcs, as well as the eastward bulges around the tips of South America, are perfect examples of this phenomenon. The alignment of the co-eval tectonic belts shows this pattern. The lineaments change directions at discrete intervals because Earth wobbles on its axis. The primary lineaments can then give a fair representation of the location of Earth's poles at the time the feature was formed.

During tectogenesis, the channels are emptied, the lithosphere overlying the channel cools, contracts, and collapses, and this creates a void in nature. As Mother Nature abhors voids, the surrounding region closes in, and compressional features such as trenches, ocean floor fabric, and fracture zones are formed.

Thus, the two opposing forces exist side-by-side. Where compression/contraction causes bathymetric highs to be juxtaposed, constriction of the channel occurs. This causes deep earthquakes as well as the multitude of shallow earthquakes.

Last, pressure increases with depth. We know this from diving. It also occurs within Earth's body as proven by artesian wells, oil wells, superdeep boreholes, and such phenomena. This pressure will tend to have an effect on the surrounding rocks, so that their molecular structure will be changed with depth, as noted in the Kola borehole. Rock porosity increases with depth. One possibility for that void is that the EMST rises to fill the void, as in the case of the Colorado dikes described above.

9

Applications

ONE THING ABOUT trying to change horses in mid-stream, it always helps if there is another horse to mount! In this case, the surge tectonic hypothesis must explain the original 29 points that the plate-tectonic hypothesis failed so miserably in trying to explain, even with all the *ad hoc* alterations and one-time instances. We will look at several outstanding items on the agenda of most normal people who are affected by these tectonic events, such as the El Nino oceanographic phenomenon.

9-1. Plateaus and Rises vs. Micro-plates

One thing about these intersections, they all underlie the rises, plateaus, and micro-plates. Where they intersect is similar to a traffic roundabout. Only so much space exists for the magma to occupy. When two or more megatrends intersect, this provides an excess amount of magma to that particular area. So, the magma, already heated, superheats the area around the intersection so that it becomes semi-molten. The magma "swirls" around depending upon the laws of whichever hemisphere the observer is in, and creates a roundabout effect. The region around the magma pool expands; it is in tension. A swirl-like pattern appears in the bathymetry/topography. This is called a vortex structure. Where the excess finds a path to the surface, it produces rise-type, positive gravity anomaly features such as seamounts, aseismic ridges, volcano arcs, and midocean ridges. These are all products of taphrogenesis where the channels are filling and flowing. The region over the channels is expanding somewhat due to the excess heat being brought to or near the surface, but this is no way to be construed as seafloor spreading or Earth expansion. Features formed at the juncture of the magma –filled channels, above the magma pools, include rises and plateaus.

Where the magma goes deeper into the mantle, this produces negative gravity anomaly vortex structures such as overlapping spreading centers and micro-plates. The negative gravity features have all been detected on the GEOSAT data and have been joined in a "vortex street" which encircles the world. Side-scan sonar images of the Easter Island and Juan Fernandez circular structures on the East Pacific Rise are replete with a system of ridges and valleys, going in a counter-clockwise direction, that look like what an orange would be if twisted by a strong hand. We have all seen tornado twisted-large oaks. They look the same in cross-section.

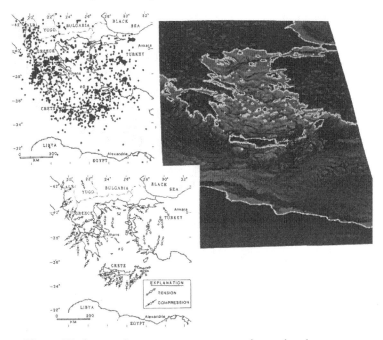

Figure 27. Aegean Sea vortex structure as shown by the earthquake epicenters (top) and the stress field (bottom).

Circle formations

One basic example lies on the northern border of the African "plate," the Aegean Sea trench/subduction zone (Figure 27). Combining the topography and bathymetry yields the picture of a circle. On investigating the earthquake pattern, the feature remains a circle. On investigating what the earthquakes are doing, those on the inside are pushing to get out (extensional), and those on the outside are pushing to get in (compression). This is typical of all downwelling vortex structures. This region is a gravity geoid low. Did we say something about a roundabout?

The region of the Ontong-Java Plateau lies in the west-central Pacific Ocean basin. It serves as an example, as it lies on the Central Pacific Megatrend. The path of the CPM is through the Van Rees and Maoke Mountains of New Guinea and along the north of the western Pacific trenches. The structural diagram for the Indonesian region clearly shows the passage of a structural feature through Irian Jaya. The southern portion of the OJP is also north of the western Pacific trenches, listed as the Vityaz Trench System (VTS) which extends over 2500 km in the form of the Manus, Kilinailau, North Solomon, Ulawan, and Cape Johnson trenches.

Presently straddling the equator, the OJP has been predicted to have formed by a South Pacific superplume between 90-150 Ma, near a ridge, and to have moved WNW to its present position from 43°S latitude and 160°W longitude. The OJP is embraced by the Udintsev and Kashima Fracture Zones. The channel which comes from the WNW and is in the center of the OJP goes out in an ENE-direction. This is the Nova – Canton Trough/Clipperton megatrend.

The next feature on the CPM is the Fiji Plateau, lying to the ESE along the Vityaz strike-slip zone. The Fiji Plateau lies atop an intersection of the Marshall-Gilbert, or Tuvalu, Megatrend, and, at 180° longitude, the Galapagos Fracture Zone appears from the ENE. The GEOSAT diagram shows five trends coming in from the ENE in the vicinity of the Galapagos Fracture Zone with a relatively clean area to the south of that and east of the Tuvalu Magatrend.

At 160°W longitude the CPM is intersected by the Manihiki Plateau. On the GEOSAT diagram there is definitely an orthogonal intersection, and this is the Emperor/Krusenstern Megatrend. The already massive

Galapagos Fracture Zone gets a fresh infusion of magma from the active Line Islands going SSE at the Tuamotu Ridge and is diverted eastwardly as the massive ridge which changes to the north fork of the Easter Fracture Zone.

This exercise can be repeated for every instance of rise-type features in oceans everywhere. For example, the Shatskiy Rise lies atop the intersection of the Mamua and Chinook megatrends. The Hess Rise lies atop the intersection of the Krusenstern/Emperor and Chinook and Mendocino megatrends. The Mid-Pacific Mountains lie atop the intersection of the Molokai and Mamua megatrends. The old Dutton Ridge plateau lies atop the Udintsev and Mendocino megatrends. The Ogasawara Plateau lies atop the North Pacific, Udintsev, Kashima, and Chinook megatrends. And so on....

Be aware that this geodynamic phenomenon also occurs in the continents, such as the previously mentioned Dasht-e-Lut in Iran. We talked about the earthquake pattern in western India. The one surge channel comes more southwesterly out of the Kashmir cusp. It meets the northerly flowing surge channel underlyning the Chagos-Laccadive Ridge in the vicinity of the Deccan traps, a rather large onshore vortex structure.

Vortex structures are easy to spot in the bathymetry, GEOSAT, and topography. Leybourne and I call it the "world-encircling vortex street."

9-2. The Surge Channel Breakout From SE Asia into the Pacific Basin

Having a stationary India for all of its history helps explain much of the tectonics of this region. A time-sequenced study based on actual rock ages shows how the present – day Pacific surge channel evolved in Southeast Asia. Sampling has given at least a Paleozoic basement to the Philippine Islands, Proterozoic granitic clasts in western New Guinea (1,250 Ma) and northwestern Irian Jaya, and on the western Malay Peninsula 1,500-1,00 Ma). With this limited information, I feel confident in stating that much of southeastern Asia is underlain by Proterozoic basement.

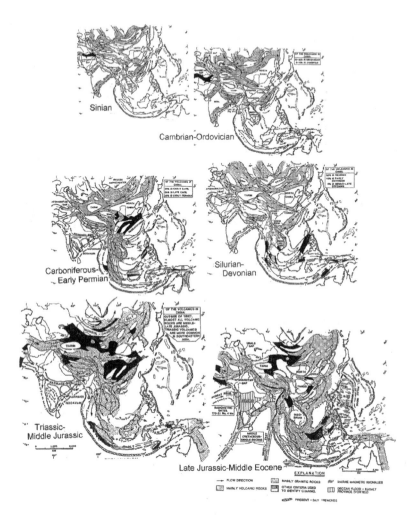

Figure 28. Surge tectonic interpretation of channel locations in SE Asia done by Art Meyerhoff. The ages of the channels has been determined by the rock ages, and the location of the channels has been determined by the location of said rocks

Many explanations exist for the evolution of the SE Asia landmass. The option here is for a far simpler explanation that leaves the mind feeling more refreshed at such a simple discovery. The Pacific surge channels arise east of the Afghanistan Gap (Figure 28) in a pinch-and-swell situation, where the Himalayan Mountains rose above the now – extinct surge channels in an orogenic phase, a period of tectogenesis. The principal continental land masses had acquired their present size and shape since the pre-Sinian, a fact which is verified by the existence of the horsetail-shaped, eastward fanning faults of central Asia (Figure 28-1) which continue through the South China Basin in the form of the Spratlys (Figure 25-12), the unnamed seamounts in the central basin, and the Paracels (Figure 25-15). Actually, depending upon the existence of old rocks in the feature, this basin, along with the features to the south, such as the North Borneo Trough (Figure 25-11), Borneo, and the Philippines themselves, all flow northeasterly. By removing the later addition of the Phil Sea "plate," (Figure 26), it could be construed as a western extension of the Chinook and Mendocino Megatrends (Figures 12, 15-3, and 15-8). Very interesting, these longer trends/lineaments, very interesting indeed!

Because the splayed surge channels were in existence for the 400 Ma preceding the end of the Ordovician (Figure 28-2), the western Pacific Benioff zones must have necessarily already been in place. During the Silurian-Devonian (Figure 28-3) the site of the Indonesian archipelago seems to have consisted primarily of deep-water troughs that surrounded shallow-water, small continental platforms. Widespread germanotypic tectogenesis and epirogenic movements took place. The trench system stretching from the Andaman-Nicobar Islands (Figure 24-12) to New Guinea was also in place. The surge channel came SE out of India in a series of five splayed channels, proving that India was in place by the Permian at the very latest.

From all structural appearances, the Banda Sea (Figure 25-2) region is an active portion of the currently inactive Indonesian vortex; a possible eddy. Sulawesi appears to be the dividing line between the two. Thus, the Banda Sea vortex is defined by the surge channel traveling eastward under the island arc and north of the trench/troughs. It is deflected north, some of which continues to the east through the Aru Trough to

Irian Jaya. Another arm goes northerly to the east of the Weber Deep (Figure 25-3), only to be deflected to the west by the arriving surge channel under the Palau-Kyushu Ridge (Figure 26-14), which intersects here at Halmahera (Figure 25-7). That portion is met by the surge channel passing along under the North and South Sula Faults. That fault system is an eastward extension of the southerly traveling Palu-Koro Fault System, and this portion completes the circle for a vortex structure. Sulawesi (Figure 25) is a K – structure, and the faults are remnant surge channels that have collapsed on themselves.

The timing of all of the surge channels flowing out of the landmass into the Philippine Basin have been delineated. The largest feeder channel for this region comes out of Sulawesi and turned north behind the Philippine Trench (Figure 25-10). This channel has been in existence for 610 Ma, or since the Sinian. Another fork welled up behind the southern Philippine Islands, a central fork passed through a gap in the trenches at 10°N, and the northern fork passed NE (Figure 25-6 and 25-7) behind the Ryukyu Trench (Figure 26-7) during the Carboniferous-to-Early Permian (362-280 Ma; Figure 28-4). The status quo remained here until the Late Jurassic (150-145 Ma; Figure 28-5). From then until the Middle Eocene (40 Ma) the surge channel formed breakouts and poured through gaps in the Benioff zones out into the Philippine Sea region (Figure 28-6). During the Cretaceous (145-65 Ma) the channels found their way south through gaps in the Benioff zone. One of these was a SE-flowing fork now called the Manila – Negros-Sulu Trench complex (Figure 25-2), the eastward pointing northern part of the engulfment zone at the Philippine Islands. Others were located at the Minami-Daito Ridge (Figure 26-15), the Kyushu-Palau Ridge (Figure 26-14), and the juncture of the Nankai Trough (Figure 26-6) and the Japan Trench (Figure 3) to wind their way southeast in the form of the South Honshu (Figure 26-12) and East (Figure 26-11) and West Mariana (Figure 26-9) Ridges. [Borehole information on the ridges gives the ages and allows the inclusion of this statement.] The Manila-Negros-Sulu Trench (Figure 25-16) complex met the N-flowing channel of the westward-pointing Philippine Trench (Figure 25-10)-East Luzon Trough system to form a vortex on Luzon Island.

E-W striking magnetic anomalies exist in the Philippine Sea basin, which are interpreted as breakout channels from the Ryukyu-Taiwan-Philippine region. The anomalies range from Early – Late Cretaceous, and they could be older. They are on strike with anomalies in the region that are believed to be as old as Carboniferous.

The history of the evolution of the flanks of the East Mariana Ridge, and active margin channel is introduced. Additionally, the western flanks and the region east of the extinct remanant arc, the West Mariana Ridge, has been called the Mariana Trough (Figure 26-10). Historically, it is listed as a backarc basin. Backarc, or interarc, basins created by extensional processes have been hypothesized since the 1960s. Backarc basins occur on the overriding plate and are always in conjunction with a subduction zone, most of which are still active and found along Pacific continental margins. Understanding the history of backarc basins is important for a number of reasons. Tectonic processes in these basins provide an analogue for the generation of oceanic lithosphere.

As the youngest of the western Pacific basins, the Mariana Trough should show the most clearly developed spreading characteristics and should be the least obscured by the overprint of sedimentation or major tectonic reorientation. This trough is in the Philippine Sea basin between 12°N and 23°N latitude. The extensional regime of tectogenesis appears to initiate in the arc structure in a progressive "unzipping" due to the presence of an active trunk channel, that point being at 23°N latitude on the South Honshu Ridge. *Great Deeps broke open* [Fountain of the]

The Mariana Trough is a region of low seismicity without great earthquakes throughout. Shallow earthquakes decrease in quantity to the south in the system, which shows that most of the active margin surge channel magma is deflected to the east as breakout channels. Paleomagnetic declination data show that the entire southern end of the Mariana arc rotated 55+-21° clockwise in the Miocene up through Guam, and that Saipan rotated 45° clockwise. The active margin channel is flowing southerly under the South Honshu Ridge and East Mariana Ridges. The eastern flow is generally blocked by the western Pacific Benioff zones except where breakout channels occur.

The Pacific Ocean basin, thought to be the oldest ocean basin, has existed for more than 600 Ma. The oldest portion of the basin is proposed to be the part adjacent to the active margin engulfment zone. The oldest borehole to date has recovered 156 Ma material (ODP Site 801). Lithosphere flexure has ascertained ages of 170-171 Ma for 12.8°N latitude, 156.8°E longitude and 21.5°N latitude, 159.2°E longitude. Davidson (1992) places the Pigafetta Basin at 160-170 Ma. A transect across a convergent margin has never been drilled down to basalt in the entire history of the DSDP and the ODP. Leg 60 at 18°N latitude attempted, but was not successful. Leg 125 also failed at 31°N latitude. Oceanic lithosphere is a 10-km thick and generally uniform shell, which is mostly basalt. As a last resort, the "M" series is still the only agent of age extant. As has been shown, the ages listed for the "M" series are not ground-truthed, so that measurement is merely used to show a chronology in relative ages, as in "this" happened before "that." However, as has been shown, continental crust underlies the ocean floor over most of the western Pacific basin east of the trenches.

Magma in the breakout channels coming off the active margin channel from the Japan Trench-Nankai Trough intersection seeped through gaps in the Mariana, Bonin, Izu, and Japan trenches to cause the Mid-Cretaceous (120-110 Ma) outpouring of volcanoes in the western Pacific basin. The bathymetric expressions of that voluminous outpouring are: (1) the Uyeda Ridge, which continues east as the Chinook Megatrend (Figure 15-3), and its partner as a K-structure, the Michelson Ridge (Figure 15-1), (2) the Dutton Ridge (Figure 15-9), which continues northeasterly as the Mendocino Megatrend (Figure 15-8), (3) the Magellan Seamounts, which continue SE as the Marshall-Gilbert trend (Figure 15-21), and (4) the Caroline Ridge on the south side of the Mariana Trench (Figure 26-3).

These anomalous ridge trends may be caused by increased amounts of magma outflow along preexisting faults. Similar magmatic outflows have been witnessed subaerially on the island of Hawaii. Given points of weakness and rupture of the lithosphere along a tectonically active area, there will be magma rising to the surface. An evolutionary scenario is suggested in which the Chinook Fracture Zone (Figure 15-3) became active before the emplacement of either the Hess (Figure 15-6) or Shatskiy (Figure 15-4) Rise. The activity in the oblique features makes it possible to explain the presence and prevalence of the 040°-striking, oblique ridges in the 060°-trending Chinook Fracture Zone. These ridges have necessarily formed at the time when differential lithosphere motion was taking place. That motion provided the mechanism for the obliqueness of the ridge trends and bathymetric signature not associated with active tears. I suggest that the rises formed after the fracture's propagation eastward; that is, after tectogenesis when the planet was on another pole-of-rotation. DSDP Site 464 on the northern Hess Rise gives an age of 98-108 Ma. DSDP Site 465 on the southern Hess Rise gives an age of 103+ Ma, making the trough at least that old. No drill holes have ever reached basalt on the Shatskiy Rise. ODP Site 810 and the geomagnetic reversals extended one Ma to 70 Ma. Isostatic compensation influenced by the rises is discounted as an agent of formation. Instead, it is merely one of propagation. The correct route this portion of the northcentral Pacific Basin has already been demonstrated for the Mendocino, Pioneer, and Murray fracture zones (Figure 15-14) and, in fact, the eastern Chinook Trough itself; both avenues are correct, and they merge in the central Shatskiy Rise. While the age of the rises remain somewhat in question, the ages of the fractures on the WSW – ENE azimuth are definitely in question.

These fracture zones/swarms were formed when the planet was on a pole of rotation other than the present pole. Their present azimuth is generally WSW – ENE. Because the primary constituents in the geomorphology are fracture valleys, the features must of necessity and by definition be older than the active surge channels. The older features are subject to reactivation, and many of them do display cross-trend seamounts and ridges within the near boundaries of those features. The younger expressions of surge channel activity are the ridges and seamount chains as has been previously noted, and their present azimuth is generally WNW – ESE.

GEOSAT data (Figure 19) and mid-plate earthquake seismicity for the central Pacific Ocean basin allow the continuation of the easterly-flowing trend from the Golden Dragon's lair, 180°, to 80°E longitude. The advantage to using the high-pass filtered GEOSAT data here is that the surge channel is only partially manifested surficially because it is still active. Also, we know that this surge channel has been active for the past 43 Ma.

In a surge-tectonic setting, the East Pacific Rise (Figure 2-6) and the San Andreas Fault (Figure 15-2), which is an almost extinct section of this trunk channel, are one in the same. The feeder channel coming from the north, which is represented by the Gorda, Endeavor, and Juan de Fuca ridges (Figure 9), intersect the older surge channels of the Chinook and Mendocino Megatrends to form the magma flood, called the Columbia River Basalts. This geostream continues to the SSE under the San Andreas Fault to go offshore as the northern portion of the East Pacific Rise. Where the East Pacific Rise bows eastward the most in the bathymetry, the location of what they term as a triple junction, is the site of a feeder channel going east. Rather than being defined as the place where the Pacific, Cocos, and Nazca plates meet (Figure 2), it is the site of the beginning of that feeder channel that will ultimately form the eastward bulge of the outer Caribbean Island arc. In the vernacular, this region is called the Galapagos Rift, and part of it is a vortex.

The next branch of the active trans-Pacific surge channel begins at the Fossa Magna region in Japan to become the active North Pacific Megatrend.

Please return to the western Pacific basin once again. For the southern portion of the Philippine Sea region, the magnetic anomalies strike N-S behind the trench system. The anomalies of eastern Indonesia strike ENE. Those of the Bismark and Solomon Seas strike northeasterly. All converge on the Caroline Basin, continue beneath the Ontong-Java Plateau, which is the site of the largest cusp in the world, and become the Central Pacific Megatrend. The channels were in existence for the 410 Ma preceding the end of the Ordovician. The channel is very active, and I will expand on that in the next section.

According to all of the material presented herein, the surge channel activity is still strong for the mid-Pacific basin. The location of that route for modern times is interesting, especially if one still believes in the plate-tectonic hypothesis. The tectonic picture of events at the western South America margin is changing rapidly. As mentioned elsewhere, the presence of paleoland during Paleozoic to Mesozoic in the Southeast Pacific near Peru-Chile Trench is in accord with the seismic interpretation. Apparently the paleoland has not been subducted and disappeared as assumed by plate tectonists, but it is still there now under the deep ocean. Investigators have defined the depth of South America's cratonal roots at 600 km, roots which are deep enough to prevent any flow from the west to pass under. The orientation of the mantle flow was shown by 350 seismograms to be parallel to the western portion of the South American craton and not to dip into the mantle. That flow is instead splitting and passing around the continental extremes. This explains why the Caribbean and Scotia arcs appear as they do.

The practicing neophyte can build on this explanation of Pacific basin surge channel activity by applying any of the methods used herein. The first criterion is an open mind, and the rest will come naturally.

But, what good is a lesson in tectonics? What good will it do anybody? Let's take a look at the next section where actual applications of the above have been, and are being, used.

9-3. Active Megatrends Affect Ocean Circulation and Climate

Despite the tremendous increase in the worldwide seismic network for earthquake monitoring, prediction of an earthquake still remains an elusive scientific goal. Until that goal is reached, is there a way that we can use this seismic data for some other, equally important and immediate societal use? Here is presented some recent approaches linking seismic and microgravity data to climate studies, thereby opening a new frontier as well as a new end-member to climate research. The idea is off the beaten track and draws upon some known but unexplained phenomenon and incorporates some of the new findings in earth and climate studies. The idea is still maturing but needs an impassionate look given the fact that (a) there is ever-increasing awareness of and concern about the waywardness of global climate, best summed up in just two words: "climatic extremes." Notwithstanding some contrary opinions regarding the climatic extremes as unusual, the two words have become dreaded for the ever-mounting losses to life and property in rather unpredictable manner and places worldwide. And, (b) with some economies still largely agriculture-based, they have high stakes in a better predictive element of climate research.

Beginning from the beginning, global climate is considered to be the resultant of interaction between ocean and atmosphere with the

Earth's three orbital parameters (eccentricity, obliquity, and precession) forcing the long-term trends (Milankovitch Series) through controlling the incoming solar radiation. Long distance correlation between sea level pressure and surface air temperature has established three global atmospheric oscillation systems (GOS): Southern Oscillation (SO) associated with the well known and much researched weather phenomenon, the El Niño, the North Pacific Oscillation (NPO) controlling North American weather patterns, and the North Atlantic Oscillation (NAO) controlling European and Siberian weather patterns. However, some interesting phenomena have been observed that raise doubt on limiting our understanding to these two end-members for understanding and predicting climate and weather patterns. For instance, increase in T-phase seismicity over hundreds of kilometers along the East Pacific Rise has been observed to precede a drop in high pressure cell of the SO, located over the Easter and Juan Fernandez Islands in the Pacific Ocean. Other studies have pointed to the increased volcanic activity and hydrothermal venting along the East Pacific Rise as well as the anomalous temperature rise in the ocean waters surrounding Indonesian archipelago – the area of low pressure cell for the SO – during El Niño phases. Repeated over seven times since 1964 until 1998 but unexplained by the current geophysical models, the increased seismicity and other tectonic phenomena along the EPR are suggested to be the "Predictors of El Niño".

A fundamental question regarding the three GOS that, to my knowledge, has never attracted the attention of meteorologists/ climatologists or any others dealing with the subject is: Why are the high pressure and the low pressure cells of the three GOS located at their specific geographic locations? Is their location determined by any of the parameters commonly thought to determine/affect global weather? Also, is it just coincidental that the wind system in the northern hemisphere vacillates generally between meridional (over North America and Europe) and zonal (over eastern Asia) flow while being largely zonal in the southern hemisphere or does it have any relationship with a broadly similar ocean ridge/vortex structure (areas of major magmatic upwelling/downwelling) orientation in the two hemispheres? And related to this, does orientation/strength of blocking systems in northern and southern hemispheres have anything to do with the orientation of vortex structures in the two hemispheres? As we know, there is nothing coincidental in nature; so, what is the natural control here?

Well, you're in luck. A group has already been trying to get funding to study such an effort. Bruce Leybourne, retired geophysicist of the Naval Oceanographic Office, and Dan Walker, retired professor at University of Hawaii Institute of Geophysics, have both tied earthquake activity to that onset, and this is roughly how it works. Bruce has been running into stone walls since he first got the idea of gravitational teleconnection in the mid-1990s. Once again the proverbial curtain has been thrown up by the "keepers of the key."

Leybourne spent many years working in the Gravity Division at NAVOCEANO. One day the apple fell on his head, and he came up with the idea of a teleconnection. Gravitational teleconnection, yet another of the 25-cent words, is an interesting concept. In order to use it, one must be able to cross the boundaries of several fields of science; to think outside the box. The basic premise was that the El Nino event might possibly be related to tectonic events. The water body displays a heated portion that is about 6-7°C warmer than the surrounding water body. That warm body bifurcates at the Galapagos Islands to pass to coastal Central America on the north and the northwestern coast of South America on the south.

Some progress has been made towards the prediction of an El Nino with the realization by Walker that earthquake swarms on the East Pacific Rise (EPR) foretell each of these events. Another study predicted that a trans-basinal feature, such as a ridge/seamount/island chain/megatrend would underlie El Nino, and this was shown to be the CPM. Leybourne hypothesized a gravitational teleconnection between the oceanographic event and tectonic events underlying El Nino. One of the delivery systems, the CPM, ends to the south of the actual El Nino event. We sought a more direct route, a more useful tectonic hypothesis to help explain the regional tectonics, and any geophysical events we could use to underline our route.

The new geodynamic model in the Pacific Basin, constructed with circulation principles common to ocean/atmosphere models, links tectonic dynamics to the El Nino phenomena with several lines of evidence:

(1) The CPM is considered a tectonic vortex street between tectonic vortices modulating the Southern Oscillation (SO) by microgravity processes. Atmospheric pressure is directly modified by internal density changes in planetary-scale (350-1200 km diameter) tectonic vortices. The low-pressure side of the SO is underlain by the Banda Sea, an upwelling upper-mantle vortex. The high-pressure side of the SO is underlain by Easter Island and Juan Fernandez rotating vortices, a set of twin downwelling upper-mantle vortices along the EPR.

(2) Another solid line of evidence was uncovered by Dan Walker when he correlated increased T-phase seismicity along the EPR near the Easter and Juan Fernandez microplates as a precursor to El Nino events. This phenomenon has occurred seven times since 1964, each corresponding to an El Nino event.

(3) Anomalous gravity trends in the mid-Pacific delineate a physical tectonic link between the Banda Sea and Easter Island/Juan Fernandez regions. The CPM is a trans-basinal feature that may provide a conduit for microgravity oscillation transfers between these regions. Gravity lineaments on high-pass-filtered GEOSAT data highlight the CPM and further delineate upper-mantle flow dynamics throughout the Pacific Basin.

(4) Although inconclusive, Bouguer gravity anomalies on the EPR found during the Mantle Electromagnetic and Tomography (MELT) experiment may indicate colder/denser mantle sinking beneath vortex structures on the ridge. This evidence is in direct conflict with the plate concept of upwelling mantle under the ridges.

(5) Microgravity increases of approximately 17 $u$gals over about a 6-month interval in early 1996 were measured in Belgium and were attributed mostly to geophysical origins. A tectonic surge of this magnitude could possibly migrate from Europe to the Pacific Basin, especially if it moves preferentially eastward like weather fronts. In this case the timing of the migration event may coincide with the 1997/98 El Nino.

(6) Additional lines of evidence indicate wavelengths filtered from the geoid data not only correspond to mantle discontinuities at

approximately 410, 660, and 1050 km, but also correspond to some planetary-scale tectonic vortex diameters.

The Easter and Juan Fernandez vortices, two counterclockwise-rotating features along the East Pacific Rise, are considered to be driven by downwelling tectonic vortices, as explained by the surge tectonic hypothesis. These twin vortices underlie the high-pressure cell of the Southern Oscillation associated with El Nino. The CPM connects planetary-scale tectonic vortices underlying the El Nino Southern Oscillation (ENSO) pressure cells. It connects the EPR across basin to the Banda Sea tectonic vortex. The Banda Sea is considered an upwelling mantle vortex underlying the low – pressure cell of ENSO.

Gravity studies undertaken in conjunction with a wide-angle seismic refraction survey during the Mantle Electromagnetic and Tomography (MELT) experiment find evidence for denser, colder mantle near a small vortex on the EPR at approximately 15° 55'S. This refutes the concept of upwelling and outpouring magma at the ridge crest, which reminds us of the humorous adage field data collectors sometimes use in reference to in-house modelers: "If the data doesn't fit the model something must be wrong with the data." Surge tectonics predicts this very action of converging mantle flow along-axis under a pressurized ridge. The ridge pressure forces denser mantle downward and volatile magmas upward in a counter-flow pattern, and this is very similar to atmospheric dynamics.

Leybourne links these lines of evidence in the theoretical framework of surge tectonics, which incorporates mantle stream flow and vortex formation processes similar to ocean/atmospheric models. The circulation pattern known as Walker circulation, which is the dominant atmospheric circulation pattern in the Pacific, can be applied directly to a Pacific Basin mantle model.

We return to the Central Pacific Megatrend (CPM). One interesting facet of this feature, especially in association with the idea of long, continuous hot lines, is that the CPM underlies the El Nino event (Figure 29). Were we to consider the possibility that water is heated from below, as on a stovetop, rather than from some alarmist idea like "global warming" due to anthropogenic causes, such as the "greenhouse effect," we might spend our tax dollars on something that may be of some use, like predicting the onset of an El Nino. Several fishermen, weather forecasters, and the like might be very interested in this.

In essence, the flow of magma along the CPM hot line/channel heats the overlying water body. The route of the CPM lies exactly beneath El Nino. The heat is produced by the magma flowing through the channels, is carried to the water body, and this heats the air above El Nino. That causes the weather pattern changes, like more hurricanes, etc., that we see during an El Nino event. This is the gravitational teleconnection at work.

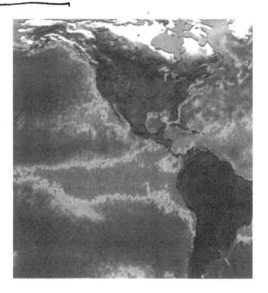

Figure 29. Sea Surface Temperature (SST) diagram of the 1997 El Nino event. The spatial relationship between El Nino and the Central Pacific Megatrend seems to be more than fortuitous.

Now, with the advent of the GPS and other information-gathering satellites, a study of Earth's dimensions has become more of a reality. If one believes that these satellites can measure in millimeters, then the following becomes easily digestible. The National Aeronautical and Space

Administration (NASA) has an ongoing program to take Gaea's measurements. In a program called GRACE (Gravity Recovery and Climate Experiment), Earth was determined to be becoming more round rather than the usual oblate shape. This is because of the ice melting at the poles as we rebound from the Ice Ages. Now, they find an anomalous movement of mass back towards the equator, so that a bulge remains in the waistline. His should not be such a surprise. In the model herein, the Central Pacific Megatrend, a large, active feature, is swelling, possibly helping the cause of El Ninos. GRACE may be a surprise to the P-Ters, but we have known about it for years. This is just one more proof of earth pulsation, an integral part of tecto – and taphrogenesis.

No problemo.

Climate being such a complex and energetic phenomenon, this new approach may not necessarily provide the ultimate all-encompassing model for climate modeling and prediction but it certainly opens up an avenue for a fresh look, particularly for isolating the natural from the human contributions to global warming. Given our vast relevant scientific infrastructure and trained manpower, we are better placed to make an early beginning in the field. If found worth its while by the experts, this topic should be taken up for future research in an interdisciplinary mode among seismologists, geodeticists, climatologists, and tectonicists – of course, those willing to look outside the box.

10

Tectonic Globaloney: Your Tax Dollars Hardly at Work

WELL, THERE YOU have it: facts vs. geophysics. Still haven't grasped the epitome of the situation? Let's look a little more closely at the disposition of your hard-earned tax dollars.

With ridge-parallel magma flow rather than ridge–perpendicular, orthogonal intersections and vortex structures; 2 Ga-old ocean floor; no convection cells because no mantle plumes or hot spots exist; a cold-core Earth; no continental drift, let alone the total disagreement of the proliferating models; no subduction; and mid-plate earthquakes, not one of the basic tenets of the plate tectonic hypothesis seems to be descriptive of the ocean basins. We have no other recourse but to abandon the constraints of that hypothesis and seek a more robust explanation elsewhere. That is not the purpose of this treatise, though, so we will leave the formulation of a valid earth geodynamic hypothesis to some aspiring young mind to digest all of these facts. We will embellish the obvious by stating what we have found.

The elevated topography, the banded earthquakes, and the elevated heat flow structures all point toward linear features. Magma channels, much the same as lava tubes, underlie the linear, heated, active features. The first assumption is that long, linear features with elevated heat flow and earthquake activity lie at the base of the lithosphere. Just as lava flows, so does the magma occupying these features, or channels as they are called.

The collective body of adherents to the plate-tectonic hypothesis raised a shield that prevented any piercing of the armor. How could such a travesty happen in this modern "enlightened" society? Who controls your tax dollars? How is it distributed? Who makes that decision? Good questions all.

Having reviewed many proposals from both the National Science Foundation (NSF) and the Office of Naval Research (ONR), I feel somewhat qualified to elaborate on this topic. They control the monies being spent by the universities on research. The Congress allots so much money per year to do this. The proposals, some of which have been running for years, are submitted to the NSF, in our case the Geosciences Department, probably Marine Geology and Geophysics Division. MG&G was budgeted $21,000,000.00 for 2002. They will send the proposals out to those whom they feel to be qualified reviewers. Several are returned, and the decision as to the amount of the allotment is made on that. A field review is conducted every two years to ascertain that the monies are being spent on what they are supposed to be spent on. All good and well.

By the way, Federal government workers, and apparently retirees, could not qualify for any of this grant money by definition. They are for private individuals/research organizations. As such, no ax is being ground here. In fact, I was honored that these erstwhile bodies considered my opinion worth having on the proposals I reviewed. Thanks, dudes.

The earthquake data listed herein and the results of drilling in the Kola Peninsula of Russia should have been enough to reassess funding earth science research. For these two earthquakes and Kola drilling brought to the fore the scandal that earth science research has become. They should have forced course correction among earth scientists from their dogmatic thinking if funding agencies were a bit more concerned about proper utilization of funds. But then, the problem is that fund managers and the big names in earth science – research and editorial boards of earth science journals – happen to be the same. That's right; they police themselves. Who is there to act as a watchdog? You may rightly ask. Nobody. Among the earth science community those who raise questions become pariah. So, the scandal goes on.

However, the monies are almost always given to the universities, and they are generally kept within the same regimes. As an example, if you had a grant for 20 years to study the blue widgets given off by passing luna moth's left wings, and you were making some progress; i.e., going to conferences, presenting papers, and generally being a good guy, the chances are very good that your grant would be renewed. Let's say that some obscure wannabe scientist discovers that the widgets are actually red, and they are part of the right wing, then your chance of getting a grant are nil under the current regime. It is not in the best interests of either the site investigators or the principal investigators to admit that they have made a mistake for all these years with all of your tax dollars.

But, that is not all.

And, still more dollars are being poured into the Chicxulub investigation. The iridium layer exists worldwide; which bolide it came from is still anybody' guess. It's not off the Yucatan, though.

How do you feel now about where your tax dollars are going? Better put your hands on your wallets/pocketbooks, friends, its gonna get worse. Globaloney is not the only shares in the Brooklyn Bridge being peddled.

One last little story: A few years ago a book called Blind Man's Bluff hit the shelves. This book was well received by the paying public according to the charts. It was also a work of fiction, purported to be based on truth. I know, because I lived one of the episodes. We on the BOWDITCH found the USS SCORPION in June 1968. We had no help from the SOSUS people. In fact, the position they gave us was 175 nautical miles off. The person claiming to have found the missing boat was also off by four months. Surely we don't need to be spoonfed yet more useless mis-information during our lifetimes. We've had a gracious plenty by now; we're sated.

ACKNOWLEDGEMENTS

I WOULD REALLY like to praise and thank my co-workers and co-authors over the years. Many have meant such a difference in my work that I was encouraged to "endeavor to persevere." In the early days the academicians who reviewed my papers and added helpful comments really made a difference, especially folks like Peter Vogt, Don Hussong, Will Sager, Rodey Batiza, and Bob Stern. Later, as my interest flagged in the early 1990s, Art Meyerhoff came along at just the right time. We kept up a steady dialog for a few years until his untimely death in 1995. He did teach me the value of real data again, gave me the insight to try the new "surge tectonic" hypothesis, and then kicked me in the rear end and told me to get back to work. Along the way I co-authored a few papers with my fellow workers, like Bob King and Bruce Leybourne, and I enjoyed my association with them, especially trying to convince King that PT had a few problems. Later I enjoyed a peaceful rapport with, and received editorial support from, Dong Choi (section 6-3), Ismail Bhat (sections 6-4 and 9-3), and Stavros Tassos (sections 6-3, 6-11, and 6-13). We have tried to keep the home fires burning in a hostile environment, but the forces of nature are too difficult to surmount. I will leave these findings to one of you out there in hopes that y'all will take the ball and run with it. Like a good Bloody Mary, this is merely an eye-opener.

And so, with that said, it's time to hit the surf.

APPENDIX I

PROFESSIONAL BACKGROUND:

EMPLOYED BY THE Ocean Survey Program of the US Naval Oceanographic Office 1966-1975, 1977-1998 – career consisted of 20 years of deep-ocean data collection (took 67 cruises and logged over 600,000 nautical miles at sea) in bathymetry, gravity, magnetics, and physical oceanography; progressing through state of the art technology of data collection from hand surveying and processing methods to full computer suite including transponders, inertial navigation, Transit and GPS satellite navigation, LORAN-C and Omega; single-beam sonar; SASS, Seabeam, and Simrad multibeam sonars; SeaMARC II side-scan sonar; sound velocity studies using Nansen casts, salinometers, Niskin samplers, and expendable bathythermographs; found missing submarine, USS SCORPION, in June 1968; was senior scientist from 1981 until 1988–30 years of office work consisted of data compilation of thousands of point charts and a couple of hundred regional charts over the years, training others including updating the training manual four times, many special projects, and publishing results; retired April 1998

Fun-in-the-Sun Things: worked on the JOIDES/USSAC Seamount Working Group for the Ocean Drilling Program; reviewed many NSF and ONR proposals during the 1980s and 1990s; was selected to get classified SASS bathymetry published in the 1970s, my part initially being guyots; moved into subduction zones and fracture zones by the mid-1980s; discovered orthogonally intersecting fracture zones in 1990, lack of deep earthquakes at subduction zones in 1991; adopted surge tectonic hypothesis for all writing in 1994; 34 feature names accepted by US Board on Geographic Names

Cruises:

1. 20 Jun-11 Nov 66 **USNS BOWDITCH** – NELant, GibStraits, and WestMed
2. 11 Jan-10 Mar 67 **USNS MICHELSON** – NWPac (Marianas region, Challenger Deep)
3. 22 Jun 67-19 Oct 67 **USNS BOWDITCH** – NELant and WestMed (Sargasso Sea, Straits of Sicily and Gibraltar, Skerki and Pantellaria Banks)
4. 15 Nov 67-13 Mar 68 **USNS BOWDITCH** – WestMed and NELant (Tagus Plain, GibStraits)
5. 26 May-12 Aug 68 **USNS BOWDITCH** – NELant (found USS SCORPION, ran transponder OPS for the USNS MIZAR, and surveyed the area)
6. 21 Jan-25 May 69 **USNS BOWDITCH** – NELant (Bay of Biscay, *Atlantic Voyageur*, GibStraits)
7. 15 Jun-25 Aug 69 **USNS BOWDITCH** – WestMed and NELant (GibStraits)
8. 26 Dec 69-26 Jan 70 **USNS DUTTON** – NELant (Faeroes, G/I/UK Gap, *Atlantic Voyageur*)
9. 25 May-31 Jul 70 **USNS DUTTON** – Arctic (Eastern Iceland through Jan Mayan Ridge, *Bluenose*)
10. 1 Feb-15 Apr 71 **USNS DUTTON** – NELant (Maury Seachannel, Charlie-Gibbs Fracture Zone, Rockall Plateau)
11. 20 Mar-12 May 72 **USNS DUTTON** – NLant (*Atlantic Voyageur*, GibStraits)
12. 10 Dec 72-18 Jan 73 **USNS DUTTON** – WestMed
13. 17 Jan-30 Mar 75 **USNS DUTTON** – NLant (surveyed TITANIC location for Bob Ballard, Azores Platform, *Atlantic Voyageur*)
14. 25 Oct-23 Dec 77 **USNS DUTTON** – NEPac (Gulf of Alaska)

15. 5 Jul-8 Sep 78 **USNS DUTTON** – NEPac (Gulf of Alaska, exploratory – *Golden Dragon*)
16. 27 Jul-6 Oct 79 **USNS DUTTON** – NPac (Emperor Seamounts – *Golden Dragon*)
17. 31 Dec 80-22 Jan 81 **USNS DUTTON** – NEPac (Gulf of Alaska)
18. 10-29 May 82 **USNS KANE** – Gulf of Mexico (Yucatan Straits)
19. 25 Jun-2 Sep 82 **USNS DUTTON** – NPac (Surveyor, Mendocino, and Murray Fracture Zones and Musicians Seamounts)
20. 26 Sep-1 Dec 83 **USNS DUTTON** – NWPac (Emperor Seamounts, Western extensions of Surveyor and Mendocino Fracture Zones – *Golden Dragon*)
21. 11 May-2 Jun 84 **R/V KANA KEOKI** – NWPac (Mariana Trough and Bonin Ridge)
22. 23 Nov-24 Dec 85 **USNS HESS** – Lant (Romanche Fracture Zone and Mid-Atlantic Ridge – *Shellback*)
23. 25 May-31 Jul 87 **USNS HESS** – NEPac (Murray Fracture Zone, Fieberling and Erben Seamounts)
24. 12 Jun-18 Aug 88 **USNS HESS** – SLant (Romanche, Charcot, and Ascension Fracture Zones, Mid-Atlantic Ridge, and exploratory lines – *Shellback*)
25. 25 Aug-24 Oct 97 **USNS SUMNER** – WPac (SChinaSea, Straits of Johor and Luzon)

Honors: Who's Who in Science and Engineering, 4th Edition (Marquis; 1998); Who's Who in Science and Engineering, 5th Edition (Marquis; 2000); Outstanding People of the Twentieth Century (International Biographical Centre; 2000)

PAPERS:

1. Smoot, N.C., 1980. Interpretation of Deep Sea Sounding Data, *Technical Papers of the American Congress of Surveying and Mapping, Fall Tech. Meeting*, pp. MS-2-D-1-10.
2. Smoot, N.C., 1981. Multi-beam sonar surveys of guyots of the

Gulf of Alaska, *Marine Geology*, Vol. 43, N. 3-4, pp. M87-M94, (also translated into Chinese and published in *Haiyang Dizhi Yu Disiji Dizhi Qingdao*).

3. Smoot, N.C., 1982. Guyots of the Mid-Emperor chain: swath mapped with multi – beam sonar, *Marine Geology*, Vol. 47, pp. 153-163.
4. Smoot, N.C., 1982. Northern Hess Rise extended by multi-beam sonar, *Tectonophysics*, Vol. 89, pp. T27-T32.
5. Smoot, N.C., 1983. Guyots of the Dutton Ridge at the Bonin/Mariana trench juncture as shown by multi-beam surveys, *Journal of Geology*, Vol. 91, pp. 211-220.
6. Smoot, N.C., 1983. Ogasawara Plateau: multi-beam sonar bathymetry and possible tectonic implications, *Journal of Geology*, Vol. 92, pp. 591-598.
7. Smoot, N.C., 1983. Ninigi and Godaigo Seamounts: Twins of the Emperor chain by multi-beam sonar, *Tectonophysics*, Vol. 98, pp. T1-T5.
8. Smoot, N.C., 1983. Detailed bathymetry of guyot summits in the North Pacific by multi-beam sonar, *Surveying and Mapping*, Vol. 43, No. 1, pp. 53-60.
9. Smoot, N.C., 1984. Multi-beam surveys of the Michelson Ridge guyots: subduction or obduction, In: Convergence and Subduction, T.W.C. Hilde and S. Uyeda, eds., *Tectonophysics*, Vol. 99, pp. 363-380.
10. Smoot, N.C., 1984. Guyots and tectonics of the Mid-Emperor chain, In: *Proceedings of the 27th International Geological Congress, Vol. 6, Geology of Ocean Basins* (VNU Science Press, Utrecht), pp. 135-152.
11. Stern, R.J., Smoot, N.C., and Rubin, M., 1984. Unzipping of the volcano arc: implications for the evolution of back-arc basins, In: R.L. Carlson and K. Kobayashi (eds), Geodynamics of backarc regions, *Tectonophysics*, Vol. 102, pp. 153-174.
12. Vogt, P.R. and Smoot, N.C., 1984. The Geisha Guyots: multi-beam bathymetry and morphometric interpretation, *Journal of Geophysical Research*, Vol. 89, No. B13, pp. 11,085-11,107.
13. Fryer, P. and Smoot, N.C., 1985. Processes of seamount

subduction in the Mariana and Izu-Bonin trenches, *Marine Geology*, Vol. 64, pp. 77-90.
14. Smoot, N.C., 1985. Guyots and seamount morphology and tectonics of the Hawaiian-Emperor elbow, *Marine Geology*, Vol. 64, pp. 203-215.
15. Smoot, N.C., and Lowrie, A., 1985. Emperor Fracture Zone morphology by multi – beam sonar, *Journal of Geology*, Vol. 93, pp. 196-204.
16. Smoot, N.C., 1985. Observations on Gulf of Alaska seamount chains by multi-beam sonar, *Tectonophysics*, Vol. 115, pp. 235-246.
17. Smoot, N.C., and Sharman, G.F., 1985. Charlie-Gibbs: a fracture zone ridge, In: G.F. Sharman, III and J. Francheteau (eds), Oceanic Lithosphere, *Tectonophysics*, Vol. 116, pp. 137-142.
18. Smoot, N.C., Delaine, K., and Gregory, R.L., 1985. A 3-D model of Nintoku Guyot to predict paleo-island morphology, *ACSM Bulletin*, pp. 23-27.
19. Lowrie, A., Smoot, N.C., and Batiza, R., 1986. Are oceanic fracture zones locked and strong or weak ?: New evidence for volcanic activity and weakness, *Geology*, Vol. 14, pp. 242-245.
20. Smoot, N.C. and Heffner, K.J., 1986. Bathymetry and possible tectonic interaction of the Uyeda Ridge with its environment, *Tectonophysics*, Vol. 124, pp. 23-36.
21. Smoot, N.C., 1986. Seamounts by SASS-chains through forearc seamounts, In: *Proceedings MDS '86, Gulf Coast Marine Technology Society*, pp. 470-477.
22. Smoot, N.C. and Richardson, D.B., 1988. Multi-beam based 3D geomorphology of the Ogasawara Plateau region, *Marine Geology*, Vol. 79, pp. 141-147.
23. Smoot, N.C., 1988. The growth rate of submarine volcanoes on the South Honshu and East Mariana ridges, *Journal of Volcanology and Geothermal Research*, Vol. 35, pp. 1-15.
24. Epp, D. and Smoot, N.C., 1989. Distribution of seamounts in the North Atlantic, *Nature*, Vol. 337, pp. 254-257.
25. Stern, R.J., Bloomer, S.H., Lin, P-N., and Smoot, N.C. 1989. Submarine arc volcanism in the southern Mariana arc as an ophiolite analogue, *Tectonophysics*, Vol. 168, pp. 151-170.

26. Smoot, N.C., 1989. The Marcus-Wake seamounts and guyots as paleofracture indicators and their relation to the Dutton Ridge, *Marine Geology*, Vol. 88, pp. 117-131.
27. Smoot, N.C., 1989. North Atlantic fracture-zone distribution and patterns shown by multibeam sonar, *Geology*, Vol. 17, pp. 1119-1122.
28. Bloomer, S.H., Stern, R.J., and Smoot, N.C., 1989. Physical volcanology of the submarine Mariana and Volcano arcs, *Bulletin of Volcanology*, Vol. 51, pp. 210-224.
29. Smoot, N.C., 1990. Mariana Trough morphology by multi-beam sonar, *Geo-Marine Letters*, Vol. 10, pp. 137-144.
30. Smoot, N.C., 1990. North Atlantic fracture-zone distribution and patterns shown by multibeam sonar (Reply), *Geology*, Vol. 18, pp. 912-914.
31. Smoot, N.C., 1991. The growth rate of submarine volcanoes on the South Honshu and East Mariana Ridges (Reply), *Journal of Volcanology and Geothermal Research*, Vol. 45, pp. 341-345.
32. Smoot, N.C., 1991. The Mariana Trench convergent margin at the Magellan Seamounts: tectonics and geomorphology, *Marine Technology Society '91 Proceedings*, Vol. 1, pp. 85-91.
33. Smoot, N.C., and King, R.E., 1992. Three-dimensional surface geomorphology of submarine landslides on NW Pacific plate guyots, *Geomorphology*, Vol. 6, pp. 151-174.
34. Smoot, N.C. and Meyerhoff, A.A., 1995. Tectonic fabric of the North Atlantic Ocean floor: speculation vs. reality, *Journal of Petroleum Geology*, Vol. 18, No. 2, pp. 207-222.
35. Smoot, N.C., 1995. Mass wasting and subaerial weathering in guyot formation: the Hawaiian and Canary Ridges as examples, *Geomorphology*, Vol. 14, pp. 29-41.
36. Smoot, N.C., 1995. The Chinook Trough: a trans-Pacific fracture zone, in: *Proceedings of the Third Thematic Conference on Remote Sensing for Marine and Coastal Environments*, Vol. II, pp. 539-550.
37. Smoot, N.C., 1997. Seafloor fabric and surge tectonics, *Proceedings of the Fourth Thematic Conference for Remote Sensing in Marine and Coastal Environments*, Vol. II, pp. 518-527.

38. Smoot, N.C., 1997. Aligned aseismic buoyant highs, across-trench deformation, clustered volcanoes, and deep earthquakes are not aligned with the current plate-tectonic theory, *Geomorphology*, Vol. 18, Nos. 3/4, pp. 199-222.
39. Smoot, N.C. and King, R.E., 1997. The Darwin Rise demise: The western Pacific guyot heights trace the trans-Pacific Mendocino Fracture Zone, *Geomorphology*, Vol. 18, Nos. 3/4, pp. 223-236.
40. Smoot, N.C., 1997. Earthquakes at convergent margins, *New Concepts in Global Tectonics Newsletter*, No. 4, pp. 10-12.
41. Smoot, N.C. and Leybourne, B.A., 1997. Vortex structures on the world-encircling vortex street: Case study of the South Adriatic basin, *Marine Technology Society Journal*, Vol. 31, No. 2, pp. 21-35.
42. Smoot, N.C., 1997. Magma floods, microplates, and orthogonal intersections, *New Concepts in Global Tectonics Newsletter*, No. 5, pp. 8-13.
43. Leybourne, B.A. and Smoot, N.C., 1997. Ocean basin structural trends based on GEOSAT altimetry data, in: Ocean Technology at Stennis Space Center: *Proceedings of the Gulf Coast Chapter Marine Technology Society*, pp. 135-140.
44. Smoot, N.C. and Murchison, R.R., 1998. Deep-ocean technology, bathymetry, and tectonics, *Proceedings of the International Symposium on New Concepts in Global Tectonics*, pp. 178-183.
45. Smoot, N.C. and Leybourne, B.A., 1998. Remotely sensed data contribute to the paradigm shift of ocean basin tectonics: the Banda Sea vortex structure as an example, *Proceedings of the International Symposium on New Concepts in Global Tectonics*, pp. 262-267.
46. Smoot, N.C., 1998. The trans-Pacific Chinook Trough megatrend, *Geomorphology*, Vol. 24, No. 4, pp. 333-351.
47. Stern, R.J. and Smoot, N.C., 1998. A bathymetric overview of the Mariana forearc. In: R.J. Stern and M. Arima (eds), Special Issue: Geophysical and Geochemical Studies of the Izu-Bonin-Mariana Arc System, *The Island Arc*, Vol. 7, No. 3, pp. 525-540.
48. Smoot, N.C., 1998. Multibeam bathymetry and the public, *New*

Concepts in Global Tectonics Newsletter, No. 8, pp. 4-8.
49. Smoot, N.C., 1998. WNW-ESE Pacific lineations, *New Concepts in Global Tectonics Newsletter*, No. 9, pp. 7-11.
50. Smoot, N.C., 1999. An appeal for using some sense, *New Concepts in Global Tectonics Newsletter*, No. 13, pp. 23-25.
51. Smoot, N.C., 1999. Orthogonal intersections of megatrends in the Mesozoic Pacific Ocean basin: a case study of the Mid-Pacific Mountains, *Geomorphology*, Vol. 30, pp. 323-356.
52. Smoot, N.C., 2000. The Darwin phoenix rises yet again. *New Concepts in Global Tectonics Newsletter*, Vol. 14, pp. 2-4.
53. Smoot, N.C., 2001. Ocean Survey Program (OSP) bathymetry history: Jousting with tectonic windmills. In: J.M. Dickins, A.K. Dubey, D.R. Choi, and Y. Fujita (eds) Special Volume on New Concepts in Global Tectonics, *Himalayan Geology*, Vol. 22, No. 1, pp. 65-80.
54. Leybourne, B.A. and Smoot, N.C., 2001. Surge hypothesis implies gravitational teleconnection of tectonics to climate: El Nino and the central Pacific geostream/jet – stream. In: J.M. Dickins, A.K. Dubey, D.R. Choi, and Y. Fujita (eds) Special Volume on New Concepts in Global Tectonics, *Himalayan Geology*, Vol. 22, No. 1, pp. 139-152.
55. Smoot, N.C. and Leybourne, B.A., 2001. The Central Pacific Megatrend. *International Geology Review*, Vol. 43, No. 4, pp. 341-365.
56. Smoot, N.C., 2001. Earth geodynamics hypotheses updated. *Journal of Scientific Exploration*, Vol. 15, No. 4, pp. 465-494.
57. Smoot, N.C., 2001. Fingernails, GPS, and Pacific basin closure. *New Concepts in Global Tectonics Newsletter*, No. 21, pp. 24-25. This one was also reprinted in *The Australian Geologist*, No. 123, June 2002.
58. Smoot, N.C. and Choi, D.R., 2003. The North Pacific Megatrend, *International Geology Review*. Vol. 45, No. 4, pp. 346-370.

ABSTRACTS AND PRESENTATIONS(*):

1. Smoot, N.C.*, 1980. Interpretation of Deep Sea Sounding Data, Niagara Falls, 8 October.
2. Smoot, N.C.*, 1981. Subduction or obduction of the Michelson Ridge, Symposium on Convergence and Subduction, Texas A&M University, 23 April.
3. Smoot, N.C.*, 1981. One pulse of the Mid-Emperor chain volcanic episodicity as swath mapped by multi-beam sonar, *Eos, Transactions, American Geophysical Union*, Vol. 62, No. 45, p. 1068, San Francisco, 8 December.
4. Vogt, P.R.*, and Smoot, N.C., 1982. Morphometric studies of guyots and other seamounts based on multi-beam bathymetry, *Eos, Transactions, American Geophysical Union*, Vol. 63, No. 45, p. 1108.
5. Smoot, N.C.*, 1982. Mariana back-arc region, *Proceedings of Symposium on Geodynamics of Back-Arc Regions*, Texas A & M University, 30 April.
6. Smoot, N.C.*, 1982. History of evolution of seamounts, *Seamount Symposium-The Origin and Evolution of Seamounts*, pp. 1-2, Lamont-Doherty Laboratory of Columbia University, 17 November.
7. Smoot, N.C.*, 1983. North Pacific fracture zones, *Proceedings of Symposium on Geodynamics of Ocean Lithosphere*, Texas A & M University, 29 April.
8. Smoot, N.C., 1983. Guyot definition altered by multi-beam sonar – an updating of paleo-bathymetry, *Science, Abstracts of 149th Meeting*, p. 150.
9. Taylor, B., and Smoot, N.C., 1983. Mass wasting of the Bonin forearc, *Eos, Transactions, American Geophysical Union*, Vol. 64, No. 45, p. 829.
10. Smoot, N.C., 1983. Guyots and tectonics of the Mid-Emperor Chain, *Tezisy, 27-y Mezhdunarodnyy Geologicheskiy Kongress*, Vol 27, No. 3, N.A. Bogdanov (ed), p. 410 (participation canceled due to Russia's downing of the Korean airliner over the Sea of Okhotsk).
11. Smoot, N.C.*, Delaine, K., and Gregory, R.L., 1984. A 3-D model

of Nintoku Guyot to predict paleo-island morphology, SEG/ USN Symposium on 3-D Marine Interpretation and Presentation, National Space Technology Laboratories, 13 March.
12. Vogt, P.R., and Smoot, N.C., 1984. Morphometry of Pacific guyots, *Eos, Transactions, American Geophysical Union*, Vol. 65, No. 16, p. 301.
13. Vogt, P.R. and Smoot, N.C., 1984. Morphology of flank rift zones of guyots and seamounts in the Pacific, *Eos, Transactions, American Geophysical Union*, Vol. 65, No. 43.
14. Vogt, P.R. and Smoot, N.C., 1984. The Geisha Guyots: multi-beam bathymetry and morphometric interpretation, *Eos, Transactions, American Geophysical Union*, Vol. 65, No. 43.
15. Smoot, N.C.*, 1984. Guyots, seamounts, and tectonics of the Hawaiian Emperor elbow by multi-beam sonar, *Proceedings Pacific Conference on Marine Technology*, OSTS/15.
16. Smoot, N.C.*, 1985. JOIDES/USSAC Workshop on carbonate banks and guyots, Scripps Institution of Oceanography, La Jolla, California, 6-8 August.
17. Vogt, P.R. and Smoot, N.C., 1985. Mesoscale multi-beam morphometry of guyots and large seamounts, *Eos, Transactions, American Geophysical Union*, Vol. 66, No. 46, p. 1080.
18. Stern, R.J., Bloomer, S., and Smoot, N.C., 1986. Submarine volcanoes in the Mariana Arc, *Eos, Transactions, American Geophysical Union*, Vol. 67, No. 16, p.411.
19. Smoot, N.C.*, 1986. Seamounts by SASS-chains through forearc seamounts, Marine Technology Society, New Orleans.
20. Smoot, N.C.*, 1986, JOIDES/USSAC Workshop on Seamounts, Lamont-Doherty of Columbia University, 4-5 June.
21. Smoot, N.C.* and Tucholke, B., 1986. Multi-beam sonar evidence for evolution of Corner Rise and Cruiser Seamount Groups, *Eos, Transactions, American Geophysical Union*, Vol. 67, No. 44, p. 1221, San Francisco, December.
22. Epp, D. and Smoot, N.C., 1988. Distribution of seamounts in the North Atlantic, *Eos, Transactions, American Geophysical Union*, Vol. 69, No. 16, p. 462.
23. Smoot, N.C.*, 1988. North Atlantic fracture valley distribution

and patterns, *Eos, Transactions, American Geophysical Union*, Vol. 69, No. 16, p. 462, Baltimore, May.
24. Bloomer, S.H., Stern, R.J., and Smoot, N.C., 1988. Size and spacing relations in submarine volcanoes in the Mariana and southern Volcano arcs: implications for the development of intraoceanic arcs, *Eos, Transactions, American Geophysical Union*, Vol. 69, No. 16, p. 506.
25. Smoot, N.C.*, 1990. Surface morphology of submarine landslides on Pacific plate seamounts. *Eos, Transactions, American Geophysical Union*, Vol. 71, p. 1578, San Francisco, 3 December.
26. Smoot, N.C.*, 1991. The Mariana Trench convergent margin at the Magellan Sea – mounts: tectonics and geomorphology, Marine Technology Society, New Orleans, 11 November.
27. Smoot, N.C.*, 1993. Geomorphic effects of seamounts in NW Pacific subduction zones, *Geological Society of America Abstracts with Programs*, Vol. 25, p. A-379, Boston, 28 October.
28. Smoot, N.C. and King, R.E., 1993. The Darwin Rise demise, *Geological Society of America Abstracts with Programs*, Vol. 25, p. A-379.
29. Smoot, N.C.*, 1994. Plate-wide Pacific trends – orthogonal fracture intersections, *Eos, Transactions, American Geophysical Union*, Vol. 75, No. 25, p. 69, Hong Kong, 29 July.
30. Smoot, N.C.*, 1994. The relation of seamounts to interplate deformation in the North Pacific, *Eos, Transactions, American Geophysical Union*, Vol. 75, No. 44, p. 582, San Francisco, 5 December.
31. Kim, J., Sager, W.W., Klaus, A., Smoot, N.C., Nakanishi, M., Khankishieva, L., and Brown, G.R., 1995. New bathymetry chart of Shatsky Rise, NW Pacific Ocean, *Eos, Transactions, American Geophysical Union*, Vol. 76, p. F329.
32. Smoot, N.C.*, 1995. The Chinook Trough: a trans-Pacific fracture zone, Seattle, 20 September.
33. Smoot, N.C.*, 1996. Real data calls for alternate tectonic hypotheses, Gulf Coast Marine Technology Society Meeting, Slidell, Louisiana, 28 February.
34. Smoot, N.C.*, 1996. Bathymetry of the Mid-Pacific Mountains:

Orthogonal fracture zone intersections and vortex structures, *Geological Society of America Abstracts with Programs,* Vol. 28, No. 7, p. A-55, Denver, 28 October.
35. Smoot, N.C.*, 1997. Seafloor fabric and surge tectonics, Orlando, Florida, 20 March.
36. Smoot, N.C.*, 1997. Ocean basin structural trends based on GEOSAT altimetry data, Stennis Space Center, Mississippi, 24 April.
37. Smoot, N.C.* and Murchison, R.R., 1998. Deep-ocean technology, bathymetry, and tectonics, Tsukuba, Japan, 23 November.
38. Smoot, N.C.* and Leybourne, B.A., 1998. Remotely sensed data contribute to the paradigm shift of ocean basin tectonics: the Banda Sea vortex structure as an example, Tsukuba, Japan, 23 November.
39. Smoot, N.C., 2001. Fracture zones, megatrends, and intersections in the Pacific. *Proceedings of the International Workshop on Global Wrench Tectonics*, pp. 1-38, Oslo, Norway, 9-11 May (also published on Internet by earthevolution.org).
40. Smoot, N.C.*, 2002. Ubiquitous megatrends obfuscate seafloor spreading concept, La Junta, Colorado, 6 May.
41. Bhat, M.I., Smoot, N.C. and Choi, D.R., 2002. Indian earthquakes shake Earth's geodynamic foundations, *Proceedings of New Concepts in Global Tectonics*, pp. 287-288, La Junta, Colorado, 8 May.

LONGER TREATISES

1. Smoot, N.C., 1986. *Bathymetric Atlas of North Pacific Guyots.* (Masters Thesis, University of Southern Mississippi, Hattiesburg), 129 p.
2. Smoot, N.C., 1991. *North Pacific Guyots,* Naval Oceanographic Office Technical Note TN 01-91, 101 p.
3. Smoot, N.C., 1993. *Bathymetry: Collection, Processing, Interpretation.* Naval Oceanographic Office Training Manual, TM 03-93, 299 p.
4. Meyerhoff, A.A., Taner, I., Morris, A.E.L., Agocs, W.B., Kamen-

Kaye, M., Bhat, M.I., Smoot, N.C., and Choi, D.R., (ed) D.M. Hull, 1996. *Surge Tectonics: A New Hypothesis of Global Dynamics* (Kluwer Academic Publishing, Dordrecht), 323 p.

5. Smoot, N.C., Choi, D.R., and Bhat, M.I., 2001. *Active Margin Geomorphology*, (X-libris Corp. Philadelphia USA), 164 p.
6. Smoot, N.C., Choi, D.R., and Bhat, M.I., 2001. *Marine Geomorphology*, (X-libris Corp., Philadelphia USA), 310 p.

Made in the USA
Columbia, SC
12 July 2018